Mark Ronan is a Professor at the University of Illinois at Chicago, and Visiting Professor of Mathematics at University College London, having held previous academic positions in Berlin, in Braunschweig, and in Birmingham where he was Mason Professor of Mathematics in the late 1980s and early 1990s. In his early career he worked on the fringes of the Classification program and knew personally all the main people involved in the modern part of this story. His work is now on geometric structures exhibiting symmetry, on which he has written numerous research papers and a textbook published by Academic Press in 1989. Besides mathematics, he has a great love for music. He has acted in more than a dozen operas at the Lyric Opera of Chicago and danced in *The Nutcracker*.

Symmetry and the Monster

One of the greatest quests of mathematics

Mark Ronan

OXFORD
UNIVERSITY PRESS

Great Clarendon Street, Oxford OX2 6DP

Oxford University Press is a department of the University of Oxford.
It furthers the University's objective of excellence in research, scholarship,
and education by publishing worldwide in

Oxford New York

Auckland Cape Town Dar es Salaam Hong Kong Karachi
Kuala Lumpur Madrid Melbourne Mexico City Nairobi
New Delhi Shanghai Taipei Toronto

With offices in

Argentina Austria Brazil Chile Czech Republic France Greece
Guatemala Hungary Italy Japan South Korea Poland Portugal
Singapore Switzerland Thailand Turkey Ukraine Vietnam

Published in the United States
by Oxford University Press Inc., New York

British Library Cataloguing in Publication Data

Data available

Library of Congress Cataloging in Publication Data

Data available

Typeset in Times
by RefineCatch Limited, Bungay, Suffolk
Printed in Great Britain
on acid-free paper by
Clays Limited, St Ives plc
ISBN 978-0-19-280723-6

5

Preface

In recent years several books on mathematics have been published, presenting intriguing pieces of the subject. This book also presents some interesting gems, but in the service of explaining one of the big quests of mathematics: the discovery and classification of all the basic building blocks for symmetry. Some mathematicians were sceptical of explaining it in a non-technical way, but others were very encouraging, and I would like to thank them. In particular I owe thanks to those mathematicians who read all, or large parts, of the manuscript: Jon Alperin, John Conway, Bernd Fischer, Bill Kantor, and Richard Weiss. I also thank my son and daughter who were always positive about the outcome, and finally my editor Latha Menon who made very helpful criticisms.

Mark Ronan,
February 2006

Contents

	Prologue	1
1	Theaetetus's Icosahedron	5
2	Galois: Death of a Genius	11
3	Irrational Solutions	27
4	Groups	42
5	Sophus Lie	53
6	Lie Groups and Physics	71
7	Going Finite	79
8	After the War	88
9	The Man from Uccle	97
10	The Big Theorem	113
11	Pandora's Box	127
12	The Leech Lattice	142
13	Fischer's Monsters	157
14	The Atlas	171
15	A Monstrous Mystery	190
16	Construction	205
17	Moonshine	216
	Notes	230
	Appendix 1: The Golden Section	238
	Appendix 2: The Witt Design	240
	Appendix 3: The Leech Lattice	242
	Appendix 4: The 26 Exceptions	244
	Glossary	248
	Index	251

To Grace Varndell,
(1909–2006)

my headmistress from primary school,
who always remembered exactly where I sat in class.

Prologue

In November 1978 an English mathematician named John McKay was reading a research paper at his home in Montreal. He worked in a branch of mathematics called group theory, which deals with the study of symmetry. It was an area that had recently produced some exceptional objects in many dimensions, but McKay was taking a break by reading a paper in number theory, the part of mathematics that deals with the whole numbers. There was no connection – or so he thought.

The largest of the exceptional symmetry objects had been called the Monster. It had not yet been constructed, but a careful examination of the data showed that the Monster – if it existed – could probably be viewed in 196,883 dimensions. Now McKay was reading about an object in number theory, and out popped the number 196,884. He was astonished. Any relationship with the Monster seemed absurd – they came from completely different parts of mathematics – but he felt he should tell someone, so he wrote a letter to John Thompson, the great guru in group theory.

Another person receiving the letter might have waved aside the coincidence as too speculative and beyond understanding, but not

Thompson. He was a sober cerebral man, and he took it seriously. He checked up on other numbers – far bigger than 196,883 – that came out of the Monster and compared them to those that emerge in number theory from the miraculous object that McKay had been reading about. Thompson found further coincidences and saw that a more detailed study was called for.

When he returned to Cambridge in December – he'd been visiting the Institute for Advanced Study in Princeton when he received McKay's letter – he mentioned these coincidences to John Conway, who had found some of the new symmetry objects himself. Conway had masses of data on the Monster, and used it to produce other sequences of numbers that might be interesting. He then visited the library and found the *same* sequences appearing in some nineteenth-century papers on number theory. He and a young mathematician named Simon Norton used these facts to make further calculations and verify that there was a definite connection between the Monster and number theory, even though we didn't understand why.

Conway dubbed the whole thing Moonshine, not because he thought it was nonsense, but because 'The stuff we were getting was not supported by logical argument. It had the feeling of mysterious moonbeams lighting up dancing Irish leprechauns. Moonshine can also refer to illicitly distilled spirits, and it seemed almost illicit to be working on this stuff.' The term soon caught on, and when I first heard it I took it to suggest something shining by reflected light, like the Moon. There may be a more primary source of illumination yet to be discovered, and this is one of the great attractions of mathematics – the deeper we go into the subject, the more there is to discover.

Mathematics will never be fully known. There will always be

deeper levels to uncover and further surprises in store. Carl Friedrich Gauss, one of the greatest mathematicians of all time, has called mathematics 'The queen of sciences', and it is a subject that compels creativity, driving mathematicians forward on quests that are beyond the power of any individual.

The quest in this book – leading eventually to the Monster and Moonshine – is to discover all the basic building blocks for symmetry. In following this quest we shall examine symmetry and how mathematicians have used it to solve deep problems. In the meantime, it is worth quoting that polymath Goethe, who wrote of symmetry:

> By the word symmetry . . . one thinks of an external relationship between pleasing parts of a whole; mostly the word is used to refer to parts arranged regularly against one another around a centre. We have . . . observed [these parts] one after the other, not always like following like, but rather a raising up from below, a strength out of weakness, a beauty out of ordinariness.*

How we go from the mathematical study of symmetry to the Monster is a long story, but I can summarize it in a few words. Most of the basic building blocks for symmetry come in one of several infinite sub-families. These sub-families combine together in larger families, but then there are the exceptions: 26 of them that don't fit in any of these families, the largest being the Monster.

Finding these infinite sub-families and then finding the exceptions is a story that takes us from France in 1830 to the 30 years following the Second World War. Showing that they form a complete list takes us right up to the present day. And finding the underlying connections between the Monster and other branches of mathematics and physics takes us into the future.

Mathematicians are explorers in an abstract world that touches the real world in unpredictable ways. More than 20 years ago when the Monster was first showing its true colours, the Princeton physicist Freeman Dyson wrote: 'I have a sneaking hope, a hope unsupported by any facts or any evidence, that sometime in the twenty-first century physicists will stumble upon the Monster group, built in some unsuspected way into the structure of the universe.'* That is how central the Monster appears to be.

Twenty years later, Richard Borcherds of Cambridge University (now at Berkeley in California) was awarded a Fields Medal for his work on Moonshine. The Fields Medal is the mathematicians' version of the Nobel Prize, though it's a rarer honour and only awarded to people under 40. Borcherds had shown that the Moonshine connections Conway and Norton established all fitted in with some new work on string theory in physics.

The fact that the Monster has connections to other parts of mathematics shows that there is something very deep going on here. No one fully understands it, and the links to particle physics are tantalizing. The Moonshine connections have spawned conferences where mathematicians and mathematical physicists meet to discuss these things, but let us begin with the study of symmetry itself, starting with the work of the ancient Greeks.

1

Theaetetus's Icosahedron

> In mathematics you don't understand things. You just get
> used to them.
>
> John von Neumann (1903–57)

In 369 BCE an Athenian philosopher named Theaetetus was
wounded in a battle at Corinth, and carried home. He contracted
dysentery and died in Athens. None of his writings survive, but
we know of his work through later commentators, and know
about him personally from Plato, who records two dialogues with
Theaetetus as the main character. One of these took place in 399
BCE when Theaetetus was still a youth, though clearly an exceptional one. Among his mathematical achievements was the classification of the five regular Platonic solids, exhibiting symmetry
in three dimensions. Here they are.

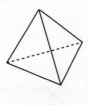

tetrahedron

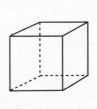

cube

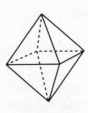

octahedron

dodecahedron   icosahedron

The Pythagoreans, that community of mystics and mathematicians founded by Pythagoras in around 500 BCE, knew about the tetrahedron, cube, and dodecahedron. The octahedron and the icosahedron are due to Theaetetus. Apart from the word 'cube', the names come from Greek and refer to the number of faces: *tetra* means four, *octa* eight, *dodeca* twelve, and *icosa* twenty.

The existence of these Platonic solids cannot be settled by making simple models, because anything you make will be imperfect. The question Theaetetus tackled is whether there is a theoretical construction in which each face is a perfect triangle, square or pentagon: all angles the same and all sides the same length. This is a question about symmetry – is there, for example, an icosahedron having perfect symmetry? This is not a trivial matter, and we shall meet the same problem later when we approach other, more complicated models of symmetry. Various sub-structures may be known, and it may seem that they should fit together to form a more complex object, but *proving* its existence can be very hard. The Monster will be a case in point.

Discovering objects that have immense symmetry is one of the later themes in this book, and the Platonic solids are good prototypes to keep in mind. Their symmetry can be described mathematically, and I want to give a rough idea of how this works. Think first in terms of mirror symmetry, by which I mean switching everything from one side of a mirror to the other. You treat the mirror as a plane, dividing space into two halves that are interchanged – like Alice interchanging with a mirror image of herself on the other side of the mirror. This is what mathematicians usually mean when they talk of reflections, or mirror symmetries.

Take the cube as an example. Take an imaginary plane that goes through the centre of the cube in such a way that each corner on one side is directly opposite a corner on the other side. Then switch everything on one side of the plane to the other. This will fix everything in the plane, but switch each point on one side with an opposite point on the other. A cube has exactly two different types of mirror symmetries. Either the plane is parallel to, and midway between, two opposite faces, or it slices diagonally through two opposite faces.

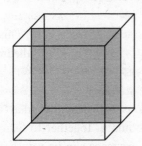 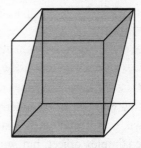

These are not the only symmetries of the cube. Several different types of rotation are also possible. For example run a spindle through the centre of two opposite faces, and turn the cube by 90° or 180°. Or put a spindle through two opposite vertices and give it a 120° turn, or through the centres of two opposite edges and make a 180° turn. You could also combine a rotation with a mirror symmetry, doing first one then the other. A cube has a great many different symmetries – how many?

The total number is 48, and they form what we call the symmetry group of the cube. Those that can be done using rotations – there are 24 of them[1] – form a sub-group. I shall call it the rotation group of the cube. The word 'group' is a technical term – it is a central concept in this book, and I will give a more precise definition later.

In the nineteenth century mathematicians found a way of deconstructing a group of symmetries into simpler groups. Those that can be deconstructed no further I shall call 'atoms of symmetry'.[2] The discovery, classification, and construction of all atoms of symmetry leads eventually to some very strange exceptions – the largest being the Monster. But if we are to gain some understanding of the Monster, we should first examine simpler situations, so let us consider the Platonic solids in more detail.

Consider the cube and the octahedron. They are deeply interconnected. Where the cube has six faces, the octahedron has six vertices, and where the cube has eight vertices the octahedron has eight faces. They both have 12 edges, but the roles of vertices and faces are switched around. This is more than just a corres-

---

[1] The reason there are 24 rotational symmetries is this: a cube has six faces any one of which can be placed on the bottom. This face can then be rotated into four different positions, and $6 \times 4 = 24$.

[2] Mathematicians call them 'simple groups', but the term 'simple' is misleading because they can be very complicated – it is used to imply they cannot be deconstructed into simpler groups.

pondence of numbers. Each one can be inscribed in the other as shown in the picture below. Place a vertex in the middle of each face of the cube, and join two of these new vertices if the faces are adjacent. This gives an octahedron, and if you do the same thing with an octahedron you get a cube. We say the cube and the octahedron are dual to one another.

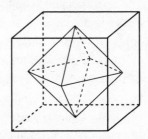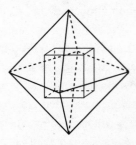

This duality between the cube and octahedron means that a symmetry of one is also a symmetry of the other – they have the same group of symmetries. As concrete objects they are different, but at the abstract level of symmetry they are the same. Abstraction is a powerful tool in mathematics; by concentrating only on certain essentials of a situation, and disregarding other aspects, one is free to pursue new results. The other aspects may have important applications, but they can be reintroduced later.

In a similar way the dodecahedron and the icosahedron are dual to one another. If you place a vertex in the middle of each face of the dodecahedron, and join two of these vertices when the faces are adjacent, then you get an icosahedron. Doing the same thing with an icosahedron yields a dodecahedron. Where the dodecahedron has 12 faces and 20 vertices, the icosahedron has 12 vertices and 20 faces. This duality means that the dodecahedron and the icosahedron have the same group of symmetries, and the same subgroup of rotations. This group of rotations has size 60, and turns out to be the smallest atom of symmetry that needs more than two dimensions. It also appears in some surprisingly different ways, and we shall meet it again later.

The word symmetry itself comes from two roots in Greek: *syn* meaning together, and *metry* referring to measurement. The idea of measuring two or more things together is obviously a useful one, and Goethe's reflections on the topic were already mentioned in the Prologue. His ideas of raising up from below, strength out of weakness, and beauty out of ordinariness have their parallels in mathematics that Goethe did not live to see. He died in 1832, as did a young mathematician named Évariste Galois, who was 62 years younger than Goethe. Galois was the first mathematician to use symmetry in solving a deep problem, and framing a new branch of mathematics. We shall meet him in the next chapter.

2

Galois: Death of a Genius

> It is important that students bring a certain ragamuffin, barefoot, irreverence to their studies; they are not here to worship what is known, but to question it.
>
> J. Bronowski, *The Ascent of Man*

In Paris on the evening of 29 May 1832 the young French mathematician Évariste Galois wrote a letter he knew would be the last of his life. It ended as follows:

> Please request publicly that Jacobi or Gauss give their opinions, not on the truth but on the importance, of these theorems.
>
> After that I hope people will be found who profit by sorting out all this mess.
>
> I embrace you with affection.
>
> É. GALOIS, 29 May 1832

There is no evidence that Karl Gustav Jacob Jacobi (1804–1851), who was an outstanding mathematician of the day, nor Carl Friedrich Gauss (1777–1855), one of the greatest of all time, ever saw Galois's letter.

The next morning, Wednesday 30 May 1832, after the sun rose, Galois lay by the side of the road, fatally shot in the abdomen. A passer-by took him to hospital. A priest was called but Galois refused to speak to him. His brother Alfred rushed to his bedside, and Évariste's last words to Alfred were, 'Don't cry. I need all my courage to die at 20.'

On 31 May his death was in all the Paris newspapers. The following extract is from *Le Précurseur*, a newspaper in Lyon:

> A deplorable duel yesterday robbed science of a young man who inspired the brightest hopes, but whose prodigious fame is only of a political nature. Young Évariste Galois ... fought a duel with an old friend, a very young man like him, and like him a member of the Société des Amis du Peuple [Society of Friends of the People] ...
>
> At point blank range, each of them was given a pistol and fired. Only one of the pistols was loaded.*

Historians of mathematics still argue about why he fought the duel, some seeing it as a matter of honour about a young lady, possibly as a set-up by agents of the police, and some seeing it as a set-up by Galois himself to go out in a blaze of glory. But though his fame as a revolutionary was transient, his mathematics was timeless: Galois theory and Galois groups are common currency in mathematics today. As a young man of 20 he joined the ranks of the immortals. How is this possible?

Évariste Galois was born on 25 October 1811 into a respectable family living in Bourg-la-Reine, a small town on the south-western outskirts of Paris. At that time, Napoleon was at the height of his power and France had achieved a stability that had been sorely lacking during the aftermath of the French Revolution. This stability was later to be lost and subsequent events had a profound and fatal impact on Galois's life.

His early years, however, were happy enough, and in 1823, aged nearly 12, he was sent to boarding school at the Lycée Louis-le-Grand in Paris. This venerable institution, founded in 1563, and renamed in the late seventeenth century by Louis-le-Grand himself (Louis XIV), still stands today in the Rue Saint Jacques – a sombre looking edifice that was recently cleaned for the first time ever. The school was a strict one: the boys rose at 5:30, silently dressed in uniforms designed by Napoleon himself, went to assembly and prayers, thence to study until 7:30, when bread and water was served for breakfast. Galois prospered in the strict regime, and in his third year, 1825–6, achieved distinctions in four subjects.

In September 1826 a new headmaster was appointed, a conservative theology teacher with rather narrow views on educational matters. He refused to allow Galois to move up to a more senior class despite his excellent results. Galois's father strenuously objected. Évariste was moved into the new class, but the headmaster eventually prevailed and he was moved out again.

This conflict between father and headmaster was part of a wider political problem. Galois's father was a liberal, and had been a staunch supporter of Napoleon. In 1815, when Évariste was nearly four years old, Napoleon returned from exile for his final 'hundred days' of power, and Galois's father became mayor of their small town. He was a popular man and retained his position when the monarchy was re-established later that year.

The new monarch, Louis XVIII, maintained an uneasy balance between liberals on one side and 'ultra monarchists' on the other, but on his death in 1824 he was replaced by his brother Charles X, whose regime was dominated by 'ultras' and supported by conservative elements in the church. The new headmaster at Louis-le-Grand had political connections to this new regime and was therefore on the opposite side of the political spectrum to Galois's father.

When Évariste was held back by the headmaster, against his father's will, the effect was devastating, and he started rejecting everything but mathematics. By the following year he had no further interest in any other subject, and at 15 he devoted all his energies to mathematics, determined to get out of the school and into the most prestigious university, the École Polytechnique, as soon as possible. He took the entrance examination in June 1828, at the age of 16, without telling his parents. This was at least a year early, and he did not succeed. This exam could be repeated once only, so everything depended on the next summer.

In the meantime a new mathematics teacher, Louis-Paul-Émile Richard, started in the autumn, and he realized immediately that he had an exceptional student on his hands. He encouraged the boy to submit an original paper to the *Annales de Mathématiques*, and it was published in April 1829. Being a man who kept abreast of the latest research, Richard was able to lead Galois in new directions, and since the boy showed such strikingly good ideas Richard tried to get him admitted to the École Polytechnique without the usual entrance examination. Unfortunately he couldn't contrive this, but he helped Galois present two written papers to the Académie des Sciences, avoiding the usual submission procedure. Richard took the manuscripts directly to Cauchy, who was a fellow of the Academy. Cauchy was an outstanding mathematician who had almost invariably only presented his own work at the Academy, so it was extraordinary that on 25 May and 1 June he presented Galois's work. In order to review these papers further the fellows of the Academy trusted Cauchy to take the manuscripts home, but he mislaid them.

Évariste Galois at the age of fifteen, drawn by a classmate.

Galois's main ideas concerned the solution of algebraic equations. Here is an example:

$$x^2 - x - 2 = 0$$

This is called an equation of degree 2 because the highest power of x is x^2 (x squared). If the highest power of x were x^3 the equation would have degree 3; if it were x^4 it would have degree 4, and so on.

The problem is to find values of x for which the equation holds true. Rather than rely on trial and error, mathematicians have a recipe for solving all equations of degree 2, shown at the bottom of the page.[1] This is called the quadratic formula (the word 'quadratic' refers to quadrature, or squaring). It is very ancient, and was first discovered by the Babylonians in about 1800 BCE, nearly 4000 years ago. They wrote in words rather than symbols, but their ancient texts, written on clay tablets, are very clear and concise.

The Babylonians even had a method for solving some special equations of degree 3 (having an x^3 term), but a general method for dealing with all equations of degree 3 had to wait nearly 3000 years until Omar Khayyám (1048–1131), the famous Persian mathematician and astronomer, devised a geometric method. He is better known for his poetry – *The Rubáiyát* – but was an

---

[1] Any equation of degree 2 can be written in the form $ax^2 + bx + c = 0$, and its two solutions are given by the formula

$$x = \frac{-b \pm \sqrt{b^2 - 4ac}}{2a}$$

The symbol $\pm$ means 'plus or minus', and the symbol $\sqrt{}$ means 'square root of'. For example, the equation $x^2 - x - 2 = 0$ in the text has $a = 1$, $b = -1$, and $c = -2$, and the formula yields

$$x = \frac{1 \pm \sqrt{1 + 8}}{2}$$

This simplifies to $(1 \pm 3)/2$, giving the two solutions $x = 2$, and $x = -1$.

excellent mathematician, and constructed the solutions to cubic equations as lengths of line segments between curves and lines.

Omar Khayyám regretted being unable to find a numerical formula, but this was finally found 400 years later during the Italian Renaissance. Modern printing was just being introduced, and this helped to accelerate the dissemination of ideas. From 1472 to 1500 over 200 new titles in mathematics were published, a huge number considering the population at the time and the small proportion of literate people. It caused a sudden upsurge in mathematics, and in the early sixteenth century four people (del Ferro, Tartaglia, Cardano, and Ferrari) moved algebra into a new era.

Scipione del Ferro (1465?–1526), a mathematics professor in Bologna, was the first to solve equations of degree 3. He never published his method, but passed it on to one of his students before he died in 1526. This student dined on the secret by challenging other mathematicians to problem solving contests where the loser paid for a series of free dinners. Since many of the problems led to equations of degree 3, the student kept winning, but in 1535 he made the mistake of challenging Tartaglia (1499–1557). (Tartaglia was a nickname meaning 'Stammerer'; his real name was Niccolo Fontana). Tartaglia made enquiries, heard that the ex-student of del Ferro had received the solution to equations of degree 3, and immediately set to work to find it himself. In two days he succeeded, won the contest, but declined the 30 free dinners. Accepting free food from a lesser mathematician was beneath his dignity, and finding a recipe for solving equations of degree 3 was reward enough!

Solving an old problem, or discovering something new, in mathematics is one of the great joys of the subject, but it can make for secrecy because you don't want anyone to know what you have discovered until you are good and ready to reveal it, with appropriate details worked out. Otherwise someone else might get the idea, work out the details themselves and claim credit. This happened to Tartaglia, as we shall see. His solution to equations of degree 3 reached the ears of Girolamo Cardano (1501–76), who had published work on medicine, astrology, astronomy, and philosophy, as well as mathematics. He asked Tartaglia for the formula but Tartaglia refused. There the matter rested for four years, but Cardano would not be denied. He cajoled Tartaglia with promises:

> I swear to you by God's holy Gospels, and as a true man of honour, not only never to publish your discoveries, if you teach me them, but I also promise you, and I pledge my faith as a true Christian, to note them down in code, so that after my death no one will be able to understand them.*

Subjected to a mix of persuasion and this protestation of faithful honesty, Tartaglia caved in. He confided to Cardano his poem for remembering the technique (this was before the days of modern notation and formulas were often given in words and remembered in poetry).

But while Tartaglia was busy with another project (translating Euclid into Italian), Cardano and his student Ludovico Ferrari (1522–65) didn't leave the matter alone. After they found out that del Ferro had obtained the solution first, and after Ferrari discovered a formula for equations of degree 4, Cardano went for publication. In 1545 he published his *Ars Magna* (*The Great Art*) and included the formula for solving cubic equations. Tartaglia was furious, and although Cardano gave credit to both Tartaglia and del Ferro, history has, rather unfairly, named it Cardano's formula.

His justification for breaking his oath to Tartaglia was the discovery that del Ferro had found it first, and as Ferrari wrote in April 1547 two years after Cardano's publication of his *Ars Magna*:

> Four years ago when Cardano was going to Florence and I accompanied him, we saw at Bologna, Hannibal Della Nave, a clever and humane man who showed us a little book in the hand of Scipione del Ferro, his father-in-law, written a long time ago, in which that discovery was elegantly and learnedly presented.*

Looking back at this period, however, it is fair to say that del Ferro, Tartaglia, Cardano, and Ferrari were all four men of genius, and as the historian of science George Sarton has written, these four 'constituted the most singular team in the whole history of science'.*

After these successes with equations of degree 3 and 4, the development stopped. It was nearly 250 years before Joseph-Louis Lagrange (1736–1813) in Berlin wrote a very influential paper with the title 'Reflections on the Algebraic Solution of Equations', which opened a new period of algebra. However, no one could find a recipe for solving equations of degree 5, or any higher degree, and in 1799 Gauss wrote, 'Since the works of many geometers left very little hope of ever arriving at the resolution of the general equation algebraically, it appears increasingly likely that this resolution is impossible and contradictory.'* That same year, Paolo Ruffini (1765–1822), a professor of clinical medicine and applied mathematics at Modena, inspired by Lagrange's work, published a 'proof' of the fact that there could be no recipe for solving equations of degree 5, or any higher degree. This was wonderful work, but his proof was long – two volumes covering 516 pages – and very hard to follow. Some mathematicians distrusted his methods and, although no one was able to refute them, the work was never fully accepted. Ruffini became dismayed at the lack of appreciation, and in 1810 he submitted a new paper on the subject to the French Academy of Sciences. The referees failed to respond in a timely manner, so Ruffini withdrew the paper, and the secretary responded politely to him:

> Your referees needed very considerable work to give their approval, or to refute your proof. You know how precious time is to realize how reluctant most geometers are to occupy themselves for a long time with the works of each other, and . . . they would have to be moved by quite a powerful motive to enter the lists against a geometer so learned and so skillful.*

Poor Ruffini. He was cracking a major problem, and was certainly on the right lines, though it now appears that his work contained an important gap. The matter was finally settled in 1824 when a young Norwegian mathematician, Niels Henrik Abel (1802–29) produced a proof independently of Ruffini's work, and it was published two years later.

Abel's paper had shown that there were equations of degree 5 whose solutions could not be extracted using square roots, cube roots, fourth and fifth roots, and so on. For some equations it *is* possible – for example, the equation $x^5 = 2$ can be solved by taking the fifth root of 2 – but the problem was to decide *which* equations could be solved in this way, and which couldn't. Abel was closing in on a method of dealing with this problem when tuberculosis carried him off in 1829 at the age of 26, but the stage was now set for the entrance of Évariste Galois, who died even younger than

Abel, and younger than any great mathematician before or since. He never reached his twenty-first birthday.

Galois measured the amount of symmetry between the various solutions to a given equation (similar to something Ruffini had done), and used it in an imaginative new way. Unfortunately in summer 1829, when Galois was still seventeen, disaster struck. Earlier in the year, a new Jesuit priest had been appointed to Galois's home town of Bourg-la-Reine. His prejudices put him in alliance with the 'ultra monarchists', and he and a member of the local administration decided to force out the mayor, Galois's father. The mayor had a penchant for writing witty rhymes that would delight members of the town council, but the clever priest now wrote scurrilous ones, under the mayor's name, making fun of some council members. The plot was successful. The elder Galois left, took his family to Paris, and committed suicide on 2 July.

Later that same month, Évariste was due to take his entrance exam to the École Polytechnique for the second time. The exam was an oral one in front of a panel of examiners, and it required a cool head. Galois was 17. His father had just been destroyed by political intrigue, and the burial service that Évariste had witnessed in Bourg-la-Reine had turned into a riot. The new priest who officiated was pelted with insults and stones and received a gash on the head. Galois's father had been a popular mayor, and the entire town later contributed to a large plaque, which stands there to this day. In the circumstances the young Galois's exam performance was not a success. One of his examiners had a skillful technique of asking simple and provocative questions. Galois lost his cool and it is said that he threw a board duster at the man. He failed the exam. No one was allowed to apply more than twice, so his failure the previous year, when he was 16, eliminated him from further consideration.

This was a disaster for him. Galois's mathematics teacher did what he could to help the boy submit a late application to a school – now known as the École Normale – that provided a two-year course. This gave a fine background for teaching mathematics, and although this was not Galois's choice, he now had no option. He started in early 1830, after making the obligatory pledge to serve the state for the next ten years.

In the meantime, Galois was anxious to have the Academy's response to the two papers he had submitted. Cauchy had taken them home, but being engrossed in his own research had failed to deal with them in a timely manner; later that year Cauchy went into political exile, and the papers were forgotten. All was not lost, however. The previous summer, the Academy had announced a prize competition, a *Grand Prix de Mathématiques*. Galois rewrote his paper and submitted it just before the deadline of 1 March. The venerable mathematician Fourier (famous for Fourier series and other essential parts of mathematical analysis) took Galois's paper home. On 16 May, Fourier died. Galois's paper was never found, and his work was not even considered.

Galois's ideas for using symmetry were profound and far-reaching, but none of this was fully understood at the time, and political events were overtaking his work. What happened was this. In August 1829 the king had appointed a new cabinet of 'ultra-ultra monarchists'. He failed to call parliament into session until March 1830, at which point the members voted to denounce the cabinet, and he responded by dissolving parliament. New elections were ordered for July 1830. These elections gave a large majority to the opposition, so the king and his ministers prepared a set of ordinances to annul them and suspend freedom of the press.

The new ordinances were made public on 26 July, and the following day there were riots. Gunsmiths' shops were pillaged, barricades were set up, but in the Rue St Jacques, Galois and his fellow students at the École Normale were reminded of their pledge to the state. From behind barred windows they could only watch in frustration while students of the École Polytechnique marched, adding insult to the injury Galois suffered from his rejection the year before.

Over the next few days, the republicans – mainly workers and students – gained control of the streets, but lacked cohesion, and those in favour of a constitutional monarchy persuaded the Duke of Orléans to enter Paris as an alternative king. On 31 July, wrapped in a tricolour flag, he was acclaimed by the crowd and accepted the regency. On 9 August he was crowned as King Louis-Phillipe.

The republicans remained a force to be reckoned with, however, and Galois was one of the most radical. At the École Normale he became an agitator, and soon started publishing diatribes against the director, whom he regarded as a cynical political opportunist. He was expelled on 9 December. By early January 1831, Galois had lost his grant, and had to give mathematics lessons to earn a living.

On 9 May 1831 at a banquet for 200 ardent republicans, Galois created a scene. With a threatening gesture and a jack-knife in one hand, he proposed a toast to King Louis-Philippe. Some guests followed his example, but others made a quick getaway. Alexandre Dumas, the well known writer, for example, escaped through a window. Galois was arrested the next day, but at his trial on 15 June he presented a clever defence in which he claimed that his actions were misinterpreted. His words were 'To Louis-Phillipe, if he betrays his oaths', the last words being drowned out by the confusion. Witnesses appeared to support this version of events or else to say they had not heard clearly because of the noise. Galois was found not guilty.

The justices did not have to wait long to catch Galois again. On Bastille Day, 14 July 1831, he was arrested wearing an illegal uniform, and carrying a knife and pistols. Three months' preventive detention was followed by a trial on 23 October, when he was found guilty and returned to prison to complete a nine-month sentence.

Galois was in the prison of Sainte Pélagie, in the section for political prisoners. He was among interesting people, one of whom was François-Vincent Raspail, a man 18 years older than Galois. He was another scientist, specializing in chemistry, who later popularized scientific knowledge, particularly in medicine, founded journals, and became an important political figure who was exiled to Belgium for ten years after the 1848 revolution. Some of his letters from prison mention Galois:

> This slender dignified child, whose brow is already creased, after only three years' study, with more than sixty years of the most profound meditation; in the name of science and virtue, let him live! In two years' time he will be Évariste Galois, the scientist! But the police do not want scientists of this calibre and temperament to exist.*

Galois's sister made frequent visits, but some of the prisoners treated him with contempt, and Raspail refers to a suicide attempt, foiled when other prisoners grabbed Galois and removed his weapon. In the spring of 1832 there was a serious cholera epidemic in Paris. Prisoners who were young or in bad health were transferred in order to lower the risks of infection, and on 16 March Galois found himself in a clinic. Here he met and fell in love with Stephanie, the daughter of the doctor in charge. She may initially have been rather attracted to him, but did not return his ardent affections, and Galois lost all hope. Rejected by the academic establishment, rejected by the state, and losing the father he loved, there remained only the republican ideals to satisfy his anger. He was released on 29 April 1832. A month later he was dead.

Historians of mathematics are not in agreement about his motives for undertaking the duel, in which he was mortally wounded, but Galois was not the only young genius to die in this way. Mikhail Lermontov, the famous Russian poet and author, also died young after being provoked into a duel at age 26, and Pushkin is another famous case, though at least he lived to the age of 37. Both these men had acquired powerful enemies, and duels were a convenient way of despatching them, but in Galois's case it is not so clear. He was so young, it is difficult to be sure what the reason was. There are different versions of the story, but what is certain is that early one morning he went to fight a duel, and was left on the field to die. Rather than give a summary of differing accounts that led up to the duel, I will follow one account, by Laura Toti-Rigatelli.*

On 7 May, Galois attended a meeting of the Friends of the People. The society had not met for several months, but a new event called them to action. The ex-queen (the wife of Charles X, who was now living abroad) had appeared in France. Her son – now 12 – was living in Prague and being educated by the mathematician Cauchy, a fervent supporter of the old dynasty, whom we met earlier in connection with Galois's first paper. The ex-queen's presence in France had thrown the royalists into confusion and presented an opportunity for action by the republicans. A pretext for a riot was needed. Galois asked to speak and said that if a body was needed it should be his. He would arrange a duel with his friend L.D. Not everyone was keen on Galois's plan, but they agreed to meet again to plan the funeral. On 29 May, Galois finalized plans with L.D. and set about writing his final letters.

On 1 June, the day following the newspaper reports of Galois's death, a meeting was held to decide tactics. At midday on 2 June, about 3000 people showed up at the cemetery of Montparnasse for Galois's funeral. They came prepared, ready to attack the police as soon as the coffin was lowered into the grave.

Police reinforcements had been placed on alert, but while the leaders of the Friends of the People gave their funeral orations, some important news was passed round. General Lamarque, appointed a Marshal of France by Napoleon, had just died. His funeral, in a few days' time, would attract an even larger crowd, more emotionally involved. A riot there would have a better chance of leading to a general uprising. A decision was taken and Galois's funeral was concluded without incident. His death at 20 achieved nothing for the revolution. For mathematics, however, his achievements will live forever.

In the cemetery of Bourg-la-Reine, the town of his birth and the place where his father was once mayor, there is no mention of Galois the revolutionary, but there is a cenotaph. It reads very simply *Évariste Galois, mathématicien, 1811–1832*.

3

Irrational Solutions

To Thales the primary question was not what do we know, but how do we know it.

Aristotle

What makes Galois's work extraordinary is that it involved such bold new ideas. This was not understood at the time of his death, but fortunately the letter he wrote the night before the duel was published, and in 1846 the well-known French mathematician Joseph Liouville republished it, with commentary. Before explaining Galois's idea, let us look again at an equation in Chapter 2 $(x^2 - x - 2 = 0)$. It factors into $(x - 2)(x + 1) = 0$, and this reduces it to two equations, $x - 2 = 0$, and $x + 1 = 0$.

Equations that cannot be reduced in this way, by splitting them into factors, are called irreducible. One example is the equation for the golden section. This ratio of one length to another keeps cropping up in art and architecture, and in nature, and is aesthetically pleasing to the human eye. It appears as early as 300 BCE, in Book 6 of Euclid, and there are many ways of describing it. Euclid used a square and a rectangle – here they are in a slightly modified form. Take a rectangular canvas having the following special proportions: if it is divided into a square at one end, and a smaller rectangle at the other, then the smaller rectangle has the

27

same proportions as the whole canvas. This can only happen when the ratio of length to width is the golden section.

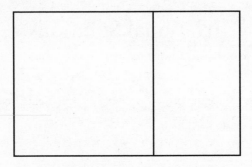

In the early Italian Renaissance, a mathematician named Luca Pacioli wrote a book on this remarkable ratio called *Divine Proportion*. The book influenced painters such as Leonardo da Vinci and Albrecht Dürer, and is evident in the work of many artists, from Renaissance painters to Neo-Impressionists like Georges Seurat and Paul Signac. The golden section cannot be written as a ratio of whole numbers, but it can be *approximated* using a remarkable sequence discovered by the medieval mathematician Leonardo of Pisa, also known as Leonardo Fibonacci. He wrote the first original book on mathematics published in Europe, in about 1200, well before the Renaissance. Leonardo had been brought up in North Africa, where he had learned the Arabic tradition of mathematics. He then visited Egypt, Syria, Greece, Sicily, and Provence before settling in Pisa. His book *Liber abaci* (book of the abacus) introduced Hindu-Arabic numerals and the place notation (units, tens, hundreds, ...) that we use today. It dealt with practical matters such as profit margins, money changing, weights and measures, and so on, but included some pure mathematics, such as the following problem.

> [You] put a pair of rabbits in a place surrounded on all sides by a wall. How many pairs of rabbits can be produced from that pair in a year if it is supposed that every month each new pair begets a new pair which from the second month on becomes productive?

This leads to a sequence of numbers in which each term (except the first two) is the sum of the two preceding it.

$$1 \quad 1 \quad 2 \quad 3 \quad 5 \quad 8 \quad 13 \quad 21 \quad 34 \quad 55 \quad 89 \quad \text{and so on}$$

This sequence – the Fibonacci sequence – appears in nature in many ways. For example the number of petals on a flower tends to be one of these numbers. Many flowers have five petals, some have only three, some have eight or 13. Daisies tend to have 21 or 34 depending on the species, and sunflowers 55.

The ratios of successive terms in the Fibonacci sequence, 13/8, 21/13, 34/21, 55/34, . . ., get increasingly close to the golden section, but never reach it. This inspired a musical creation by the Norwegian composer Per Nørgaard in his composition *Voyage into the Golden Screen*. In the first part of this work he uses beat frequencies in the Fibonacci sequence to give ratios of beats that approach the golden section. In the second part the listener has passed through a 'golden screen', and hears harmonious melodies carried by strings and woodwind.

The precise value of the golden section can be calculated using the following equation, where x is the golden section (see Appendix 1):

$$x^2 - x - 1 = 0$$

29

This equation does not split into two factors. It is irreducible, and has two solutions: $(1 + \sqrt{5})/2$ and $(1 - \sqrt{5})/2$. The golden section is the one having a plus sign for the square root, and it works out to be approximately 1.618. Neither solution is a ratio of two whole numbers, so we call them irrational – this doesn't mean unreasonable or illogical, but simply comes from the word *ratio*. A number that *can* be written as a ratio of two whole numbers is called rational.

The first people to notice the existence of irrational numbers were the Pythagoreans – the followers of Pythagoras who lived in Croton on the southern heel of Italy. They wished to express the universe in terms of whole numbers and ratios of whole numbers, and it disconcerted them that the ratio between the diagonal of a square and the length of one of its sides is irrational. It upset the idea that nature, like musical harmony, should be based on ratios of whole numbers. The brotherhood was troubled and one member was expelled when he wanted to broadcast this new knowledge. An apocryphal story even has him being drowned at sea to silence him. However, the existence of irrational numbers became a well-known fact.

In the equation above, the two solutions can be interchanged by switching the sign of the square root. Interchanging irrational solutions is exactly what Galois was doing and by examining the group of allowable interchanges, he could detect whether or not the solutions to a given equation could be expressed in terms of square roots, cube roots, and so on. We will come back to this later.

Interchanging things is an old trick, beloved by magicians. When I was a child my father used to do conjuring tricks, one of which illustrates the point rather well. There were two wooden stands, each supporting a flat wooden rabbit, one white and one black. He placed a green cover over each rabbit, mumbled a magical incantation and the rabbits appeared to change places. The white one on the left was now on the right, and the black one on the right was now on the left. The trick was repeated several times and the rabbits kept changing places.

After several repetitions the trick seemed to become obvious. The rabbits were not really changing places, it was just that the stands and their covers were being turned round. Each rabbit was both black and white, black on one side and white on the other. But this was a sucker trick, one that you think you understand until suddenly at the end you find you have been fooled. Finally both covers were removed at the same time, showing one rabbit to be yellow and the other red!

The relevance to Galois's work is that the black and white rabbits are like the solutions to an irreducible quadratic equation. If one solution is $(1 + \sqrt{5})/2$ the other solution *must* be $(1 - \sqrt{5})/2$; they are like two sides of the same thing, hidden from view by the equation itself.

Galois, of course, was concerned with equations of a higher degree where there are more than two solutions. So he was dealing with rabbits that could appear not just black or white, but also red, yellow, and other colours. The more colours, the more possible interchanges, and the complexity of the interchanges was what Galois was studying. Eventually this complexity renders a general formula for equations of degree 5 or more impossible.

Galois knew that the solutions to an irreducible equation must be irrational, and the number of solutions equals the degree of the equation. A quadratic equation (one of degree 2) has two solutions, a cubic equation (one of degree 3) has three solutions, and so on. This follows from the 'fundamental theorem of algebra', first proved in 1815 by Gauss.

Irrational solutions, like the quarks of quantum physics, only appear in multiples – and in each multiple there is a symmetry between them. Galois's genius was to analyse this symmetry, rather than the solutions themselves, which he could treat as objects, like rabbits, that could be interchanged with one another.

Interchanging several things at once is called a *permutation*. This is a mathematical term meaning a rearrangement of objects, such as beads on a string, or people sitting round a table. Sometimes it refers to an alternative arrangement, as in 'there are six permutations of three beads on a string', and sometimes it refers to the act of rearranging, as in 'the permutation interchanges the end two beads, leaving the others as they are'. The act, or operation, of rearranging is what we need.

Suppose, for example, that you have three chairs around a table occupied by Anthony, Beatrix, and Charles, in a clockwise order.

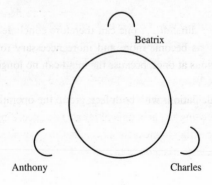

Beatrix

Anthony   Charles

First interchange Beatrix with Anthony, leaving Charles in place; that's a permutation. Then interchange Beatrix with Charles, leaving Anthony in place; that's a second permutation. The first permutation followed by the second yields a third permutation: it sends each person to the seat on their left.

What Galois examined was a system of permutations having the property that one followed by another always yields a third permutation in the same system. He called such a system a *group*. Groups arise naturally when you permute similar parts of a fixed pattern. If two permutations both preserve the pattern, then so does the permutation obtained from one followed by the other.

Galois's group of permutations for a given equation allowed him to ignore the technical details of how the solutions could be expressed. By concentrating on the possible ways of permuting them among themselves, he was able to take an eagle's eye view, and this is the magic of mathematics – allowing us to avoid technicalities and concentrate on the main game. As he wrote in an unpublished preface to his work:

> Since Euler [a mathematician who calculated things effortlessly and could write a whole paper between the first and second calls to dinner], calculations have become more and more necessary but more and more difficult . . . one can therefore conclude with certainty that it has become more and more necessary to embrace several operations at once, because the mind can no longer stop to look at details.
>
> Jump on calculations with both feet; group the operations, classify them according to their difficulty and not according to their form; such according to me is the task of future geometers; such is the path I have embarked on in this work.*

For a given equation, Galois grouped all the allowable permutations together. This is now called the Galois group of the equation. It has become a central part of mathematics, going beyond the solution of algebraic equations to play a vital role in modern number theory.

In Galois's work a vital component was the idea of deconstructing a group into simpler groups. When this process is taken to a conclusion, one reaches groups that can be deconstructed no further. This is rather like deconstructing more familiar objects. For example, a car can be deconstructed into a great many components, all carefully listed in the parts manual. Some are very simple: nuts and bolts, for example. Others are more complicated: pistons, engine block, and so on. With groups of permutations the really simple components are the *prime cyclic groups*.

Take an operation – such as a rotation, or a permutation – and keep doing it until everything is back where it started. The number of times you must do the operation to achieve this is called its *order*, so, for example, a mirror symmetry has order 2, and a rotation by 90° has order 4. A group generated by a single operation is called *cyclic*, and its size is the order of the operation you started with. A cyclic group of size 2, for instance, is generated by an operation of order 2. The nature of the operation is not important, and an operation of order 2 could take many forms: a mirror symmetry, a rotation by 180°, a permutation switching a pair of objects, and so on. We don't worry about what it *could* be, but simply treat it abstractly without regard to the various ways it might be represented. This is what group theory is all about. You take a group of operations, but tacitly ignore the way in which they first arose, because they could have arisen in a different way. When the order of an operation is a prime number it generates a *prime* cyclic group.

Cyclic groups are basic, but the prime cyclic ones are the most basic of all. For each prime number *p* (2, 3, 5, 7, 11, and so on) there is exactly one cyclic group of that size. Many groups can be deconstructed into prime cyclic groups – but not all. The distinction between those that can and those that cannot is critical in Galois's work. He obtained a group of permutations from an equation, and the point was to deconstruct it into simpler groups, as far as possible. As he said in the letter he wrote the night before the fatal duel: 'If each of these groups has a prime number of permutations, then the equation will be solved using roots [meaning square roots, cube roots, etc.]; otherwise, not.'* In other words, when Galois's group for a given equation could be deconstructed into prime cyclic groups, then the solutions to that equation could be expressed in terms of square roots, cube roots, and so on.

This led to an interesting phenomenon. According to the fundamental theorem of algebra, every equation has solutions. According to Ruffini and Abel there are equations of degree 5 whose solutions cannot be expressed in terms of square roots, cube roots, and so on. So the conclusion was inescapable. For some equations Galois's group of permutations could not be deconstructed into prime cyclic groups. Apparently the prime cyclic groups were not the *only* atoms of symmetry, and as Galois wrote, 'The smallest number of permutations an [atom of symmetry] can have, when this number is not prime, is 60.'*

This group of size 60 is the group of *even* permutations of five objects, and is the first in a sequence of symmetry atoms that goes on forever. But what is an even permutation, as opposed to an odd one?

To illustrate this distinction, here is a puzzle, invented in the nineteenth century by Sam Loyd (1841–1911), the USA's greatest puzzle master. It consists of a small frame containing 15 tiles that can slide in a four by four array; there are 16 positions, one of which is blank. Sometimes the tiles are labelled by numbers, sometimes by letters, and sometimes they are painted with parts of a picture that has to be recaptured after the tiles have been scrambled by a series of random moves. Each move interchanges the blank space with a neighbouring tile.

In Loyd's original version, the tiles were numbered from 1 to 15 and were placed in an almost correct order, except that the last two tiles, 14 and 15, were transposed. The challenge was to get them into the correct order with the blank space back in the bottom right hand corner. Sam Loyd shrewdly offered a reward of $1,000 (worth well over $100,000 today) for a solution to this problem. His money was safe, however, because the problem is impossible!

1	2	3	4
5	6	7	8
9	10	11	12
13	15	14	

This impossibility is due to the difference between even and odd permutations. Here is the idea. Forget about the puzzle for a minute, and just think about permutations. A permutation that

36

transposes the position of two objects, leaving everything else alone, is called a transposition. For example, if six people were sitting round a table and two of them swapped places while everyone else remained seated, that would be a transposition. Using a suitable sequence of transpositions, any desired permutation can be achieved.[1] But here is the extraordinary thing: if you can do a permutation using an *even* number of transpositions, then you cannot do it using an *odd* number, and vice versa. Every permutation is either even or odd in this sense, but cannot be both. For example, if you change the order of chairs round a table by a sequence of seven transpositions, you have done an odd permutation. If someone says they can do the same thing using exactly six, you can bet money they are wrong. They *might* be able to do it in five, but not in six. Your permutation was odd and they cannot do it using an even number of transpositions, no matter how hard they try.

Now back to the puzzle. The interchange of tiles 14 and 15 is a single transposition, and therefore an odd permutation. But any permutation that replaces the blank space at the bottom right must be even for the following reason. Each move transposes the blank with a neighbouring tile. If you imagine the 16 spaces being coloured white and black like a chess board, then each move sends

[1] Think of rearranging a collection of people sitting round a table. Take one person A, whom you wish to be in the seat presently occupied by B. Interchange A with B, leaving everyone else in place. Both A and B were in the wrong seats, but A is now in the correct seat, so the number of people in correct seats has increased: by one if B is still in the wrong seat; or by two if B is now in the right seat. Continue with further transpositions if necessary until eventually everyone is in the correct seat. For example, with six people round a table, any permutation can be achieved by at most five transpositions.

the blank from white to black, or black to white. An odd number of moves will change the colour of the blank space, so if it ends up where it started the number of moves is even. No matter what moves you do, if the blank space returns to the bottom right corner you have done an even permutation. You cannot end up transposing two tiles, and returning everything else to its original position, because that is an odd permutation. Sam Loyd's cash offer for a solution was safe, and he knew it!

Anyone who has tried this puzzle knows it needs some thought. If you simply make a series of random moves and hope for the best, there is little chance of hitting on the right arrangement because there are 10,461,394,944,000 possible patterns. This is the number of even permutations of 16 objects (in this case 15 tiles and one blank space).

You can calculate this by first working out the total number of permutations of 16 objects. Think of 16 beads on a string – it doesn't matter what the objects are – and string them from left to right. There are 16 choices for the first bead, 15 choices for the second, 14 choices for the third, and so on. The total number of arrangements is therefore $16 \times 15 \times 14 \times \ldots \times 2 \times 1$, which is 20,922,789,888,000. This is the number of permutations of 16 objects. Half these permutations are even and half are odd, so we divide by 2 and find that the number of even permutations is 10,461,394,944,000.

The reason for dwelling on even permutations is that whenever five or more objects are being permuted, the group of all even permutations is 'simple' – it is an atom of symmetry. In Galois's work an equation of degree 5 has five solutions, and for many such equations his group of permutations contains this 'simple' chunk, so it cannot be deconstructed into prime cyclic groups. This implies that the solutions cannot be reduced to a combination of square roots, cube roots, and so on, and hence there is no formula for equations of degree 5 or more. It is an elegant way of seeing that a formula for solving such equations is not possible, and is a good illustration of Aristotle's quotation at the beginning of this chapter that it is not what we know, but how we know it.

The group of even permutations of five or more objects is an atom of symmetry – a group that cannot be deconstructed, and is not prime cyclic. As the number of objects increases these groups grow exponentially. They get increasingly large and complex as more objects are being permuted. Here are the sizes of the first few:

No. of objects	5	6	7	8	9	10
No. of even permutations	60	360	2,520	20,160	181,440	1,814,400

In this book we are looking for the atoms of symmetry – that's how we shall uncover the Monster – and we now have a whole series of them. But the groups of all even permutations get too large too quickly. They are like a sequence of increasingly gigantic worlds, but inside they contain some fascinating objects. As an analogy think of the planet Earth. The Earth cannot be deconstructed and put together again, but from close up it contains interesting objects like trees. They too cannot be deconstructed, and they in turn contain things such as leaves. And the leaves contain cells, and the cells contain complex molecules, and the molecules contain atoms.

Now to the atoms of symmetry. Most live in predictable families, but others have astonished mathematicians. Later in the book, for example, we shall meet two exceptional symmetry atoms, each permuting 100 objects. One has size 604,800, the other 44,352,000. These may seem large but they are tiny compared to the world they live in. They are both subgroups of the group of even permutations of 100 objects, which has the following size:

46,663,107,721,972,076,340,849,619,428,133,350,245,357,984,132,190,
810,734,296,481,947,608,799,996,614,957,804,470,731,988,078,259,
143,126,848,960,413,511,879,125,592,605,458,432,000,000,000,000,
000,000,000,000

This is so gigantic that even the larger of the two symmetry atoms above (the one of size 44,352,000) is less than one million trillion trillion trillion trillion trillion trillion trillion trillion trillion trillion trillion trillionth the size of it.[2] It is impossible to grasp the immensity of this, but an analogy may help: this ratio is more than a trillion trillion times larger than the ratio of the volume of the whole visible universe to the volume of a single atom.

The question is how we are to find these mysterious atoms of symmetry that exist somewhere in the universe of all permutations. Finding them is far harder than looking for the proverbial needle in a haystack because one is searching for things one doesn't know about in a universe so immense that it is beyond comprehension.

---

[2] I am using the terms trillion and billion in their American usage as 1,000,000,000,000 and 1,000,000,000. In European usage billion means million squared, which is 1,000,000,000,000; and trillion means million cubed, which is 1,000,000,000,000,000,000.

Bear in mind that we are looking for 'simple' groups – the atoms of symmetry. Any mathematician can construct groups that are not 'simple' by putting simpler groups together, like using a construction kit to build complex objects from simple ones. Some of these constructions are quite subtle; for example, you can combine four groups of size 2 in fourteen different ways to obtain a group of size 16. Evidently there are lots of different groups out there, but the 'simple' ones – the atoms of symmetry – are far rarer. Here is a list showing the size of those having fewer than 2,000 operations.

$$60 \quad 168 \quad 360 \quad 504 \quad 660 \quad 1092$$

The one of size 60, and that of size 360, are the groups of all even permutations of five objects and of six objects respectively. The others in the list above, like the elements of chemistry, fit into a 'periodic table'. More on that later.

4

Groups

The chief forms of beauty are order and symmetry and
definiteness, which the mathematical sciences demonstrate
in a special degree.

Aristotle

In the mid-nineteenth century the idea of a group was still quite
new, and the first methods for finding 'simple' ones were to look
at groups of permutations. Other methods came into play later, but
let us start with permutations. The idea is to restrict the permutations in various ways to obtain sub-groups. Here is an example.

Four people sit down to play bridge. There are eight ways of
arranging the seating so that the two bridge partnerships are
undisturbed, and they are shown in the following diagram.

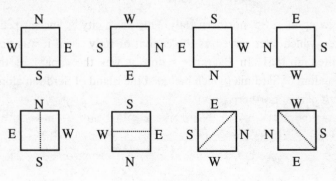

From the arrangement in the top left-hand corner, the other arrangements are obtained by rotating (top row), or by interchanging positions across the dotted lines (bottom row). These operations form a group in Galois's sense: one operation followed by another gives a third operation in the same group. For example, a clockwise rotation by 90° followed by a left/right flip takes you from the first position to the second position and then to the last position in the bottom right. This combination yields the same result as a diagonal flip, but if you do the left/right flip first and the 90° clockwise rotation second, you get the other diagonal flip. The order in which two operations are performed makes a difference to the result.

The group of *all* permutations of four people has size 24, but by restricting to permutations that preserve the bridge partnerships, we obtain a sub-group of size 8. Notice that 8 is a divisor of 24, and this is a general fact: if one group is a sub-group of a larger group then the size of the smaller divides the size of the larger. This is called Lagrange's Theorem, named in honour of Lagrange, whose work on algebraic equations was already mentioned and had a major influence on Galois.

Lagrange was born Giuseppe Lodovico Langrangia in 1736, and grew up in Turin, northern Italy. This was a city of ancient pedigree, which later served as the capital of Italy when it was first united in 1861. In Lagrange's time it was the capital of the kingdom of Sardinia (which included the island of Sardinia along with part of northern Italy).

Lagrange's father worked in the treasury, but lost most of his money in speculation, and Lagrange later said, 'Had I been rich, I would probably not have devoted myself to mathematics.' As it was, he did, and by the time he was 30, Frederick the Great of Prussia offered him a chair in Berlin. Frederick (called the philosopher king by Voltaire) wrote to Lagrange saying that as 'the greatest king in Europe' he wished to have 'the greatest mathematician in Europe' at his court. The job had excellent working conditions and salary, and Lagrange was delighted. He married and moved to Berlin, where he produced no end of papers on mathematical subjects from number theory to the stability of the solar system. Berlin is where he wrote his paper on algebraic equations, which inspired all the work that followed, culminating in that of Galois.

After more than 20 years in Berlin, Lagrange moved. His wife had died, his patron Frederick the Great had died, and Louis XVI of France invited Lagrange to Paris. He moved there in 1787, took up apartments in the Louvre, and soon married again, to a much younger woman – daughter of a well-known astronomer.

Within two years of his arrival in Paris the French Revolution started, but Lagrange avoided all factions and political entanglements, just as he had done in Berlin. He remained safe, even during the reign of terror, unlike some others. When the great chemist Antoine-Laurent de Lavoisier was guillotined, Lagrange commented, 'It required only a moment to sever that head, and perhaps a century will not be sufficient to produce another like it.' Lagrange died in 1813, aged 77, by which time Napoleon had made him a senator and a Count of the Empire.

Lagrange's work on permutations is only a tiny part of his output, but his theorem was a vital one. Theorems are the lifeblood of mathematics. A theorem is a statement that has been proved true; without them we are lost because we cannot be certain that we are building on a firm basis. If something *looks* as if it is true, and we build it into our theory, but find later that it is false, then part of the edifice collapses. Later results will have been proved by assuming the truth of the false result, and they will then have to be completely re-proved. Mathematicians are very careful about this. An important result, one that will be used elsewhere, has to be proved to everyone's satisfaction.

Theorems are essential, and this is how mathematics makes progress. It is rather different from theoretical physics in this sense. As the famous physicist Richard Feynman said, 'The whole purpose in physics is to work out a number, with decimal points, etc.! Otherwise you haven't done anything.'* Well, in mathematics, the whole point is to state and prove a theorem. Some theorems, of course, are more important than others, and most are rather specialized results that map out the mathematical landscape. If part of that landscape ceases to be of concern to mathematicians its theorems may simply gather dust in the bowels of university libraries, but some results are of perennial relevance, and many of the early Greek theorems in Euclidean geometry are fine examples.

Lagrange's theorem leads to an interesting question. Suppose you have a group of size 60. It could easily have a sub-group of size 15 because 15 is a divisor of 60, but *must* it contain a sub-group of this size? The answer is no in general, but is yes in an important special case. If the divisor is a *prime* number then there is a sub-group having that size. For example, a group of 60 elements must contain sub-groups with two elements, three elements, and five elements, because 2, 3, and 5 are prime numbers that are divisors of 60.

This was proved in 1845 by Augustin Cauchy. He has appeared already in connection with Galois, and was a leading light in the mathematical world and a mathematician of wide-ranging interests. If he was sent a paper to referee, he was quite likely to try to improve the results and publish a new work himself. For example, on 1 March 1847, Gabriel Lamé published a proof of Fermat's Last Theorem (a famous conjecture that was already 200 years old).* Lamé's proof rested on unproved assumptions, and for several weeks Cauchy published notes attempting to validate these assumptions. Then on 24 May the German mathematician Ernst Kummer produced a counter-example, showing that Lamé's assumptions were wrong (and hence his proof of Fermat's Last Theorem was invalid). One might suppose Cauchy would remain silent, but not at all. Two weeks later he presented results generalizing those of Kummer.

Cauchy was an astonishingly productive mathematician and wrote research papers at a terrific pace. French mathematics had, and still has, a regular bulletin called the *Comptes Rendus*, in which research notes are published very swiftly. In a period of less than 20 years, Cauchy published 589 notes in this bulletin, and submitted yet more for which there was no space. This is in addition to the more than 800 research papers he published.

As well as being an extraordinarily active mathematician, Cauchy was sharp-witted, opinionated, a religious conservative, and a staunch monarchist. He was on the opposite side of the political spectrum from Galois (who was sharp-witted, opinionated, and a fervent republican). But they had rather different upbringings. Cauchy was born in 1789, the year of the French Revolution, and his family fled Paris for the French countryside when the Reign of Terror started in 1793. He developed an abiding dislike for revolution and republicanism, and became a supporter of religious values and the repressive monarchy of Charles X. But behind this reactionary attitude there was a man of principle. He disapproved of Louis-Phillipe, who became king of France after the revolution of 1830, and he went into exile rather than take the oath of allegiance. This was an odd thing to do since he was under no pressure to do anything other than take the oath, but principles were principles, and he went first to Fribourg, thence to a position in Turin, and later to Prague. He returned in 1838, but could not reclaim his university employment because he had not taken the oath, so he accepted a position with the Academy of Sciences instead. In 1848 the Second Republic was formed, the oath of allegiance was abolished, and Cauchy returned to his university position. Four years later in 1852 the oath was reinstated, but two people were exempted, by the grace of the emperor Napoleon III. One was Cauchy (the other was a famous physicist named Arago).

Cauchy sounds like an awkward man, but in truth there was another side to him. He was a devout Catholic and played a leading role in several charities, including one for unwed mothers and another doing rescue work for criminals. At one point he spent his entire salary for the poor of the small town near Paris where he lived. When the mayor begged him to spare himself some money he responded, 'Don't worry, it's only my salary; it's not my money, it's the emperor's.'

In early May 1857, Cauchy sent yet another paper to the Academy, promising further results in another few weeks, but on 22 May he died.

With results like those of Lagrange and Cauchy, along with

Galois's deep ideas, there was clearly a need for groups of permutations to receive a more sustained and systematic treatment, and the person who rose to the occasion was Camille Jordan. He was an engineer by profession, like his father, but became a professor of mathematics in Paris. His work covered a wide spectrum, and he wrote the definitive text on analysis, a branch of mathematics dealing with quantities that can change gradually from one value to another. In 1870 he published his *Traité des Substitutions* (*Treatise on Permutations*), which became the standard reference for group theory for the next 30 years.

Jordan was 32 when he published his *Treatise*, and he lived another 52 years, surviving four of his sons who died in the First World War (1914–18). He outlived his wife, and sadly only three of his eight children were alive when he died in 1922, aged 84. But Jordan's work gave group theory a foundation on which future generations of mathematicians could build. His exposition attracted a wide audience and his fame spread well beyond France. Foreign students attended his lectures and two among them, Felix Klein from Germany and Sophus Lie from Norway, later produced ideas leading group theory in new directions. We shall meet them both soon.

In his *Treatise on Permutations*, Jordan explained Galois's work and showed how to deconstruct any finite group into simpler groups. This doesn't mean finding sub-groups; it means finding two or more groups that can be combined to form the original group. At least one of these must be a sub-group of the original, but a group can be built in layers, like a cake. The layer at the bottom is a sub-group, but another group might piggy-back on top of it, like icing on a cake.

The idea of breaking things up into simpler components is

pretty fundamental in science, and the trick is to reach a stage where the components are as simple as possible. For example, a physical substance can be broken into molecules, and these in turn may be split into even simpler molecules, but the process comes to an end when you reach the level of atoms. At that stage it doesn't matter how you did the decomposition, you always get the same collection of atoms. A result of Jordan, later extended by a German mathematician, showed it was the same with groups – no matter how you do the deconstruction you always get the same collection of 'simple' groups, the same atoms of symmetry.

Most large groups can be deconstructed into simpler components, and the group of symmetries for the Rubik cube provides a good example. As you rotate each face of the cube by 90°, you permute the corner pieces, and the side pieces among themselves. The group of permutations generated by all sequences of such rotations has size greater than twenty million million million, but it can be deconstructed into the following atoms of symmetry: the group of all even permutations of the eight corners, the group of all even permutations of the 12 edges, and prime cyclic groups of sizes 2 and 3 that have the effect of flipping or rotating the edge and corner pieces. The Rubik cube is a hard puzzle because its symmetry group is so large, but the deconstruction of this group makes it possible to come up with a reasonable technique for solving it.

Now consider the symmetries of a solid cube, as in Chapter 1. At that point there was no mention of permutations, but any group of symmetries *can* be treated as a group of permutations, and the cube is a good example.

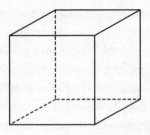

A symmetry of the cube permutes the eight corners among themselves. But while the corners are being permuted, so are the edges and faces of the cube, so the symmetry group of the cube may be thought of as a group of permutations in several ways: on its eight corners, or on its six faces, or on its 12 edges.

This makes things interesting, and more complicated, but mathematicians have a nice way of avoiding such complications by stepping back into a world of greater abstraction. They do this by studying *groups* in the abstract. These groups can reveal themselves as groups of permutations, or groups of motions, or groups of transformations of one sort or another. But they can be created and studied in the abstract, and this is exactly how the Monster was discovered.

The Monster did not appear as a group of permutations, though it can certainly be represented in that way, nor did it first reveal itself as a group of symmetries, though it can certainly be seen in that way too. It first arose as an enormous collection of operations: something to be studied, something to be constructed (did it really exist?), something to be understood. The fact that mathematicians can get a vague but increasingly precise view of something that is 'out there' may come as a surprise to some people. We are not usually thought of as creative artists, yet in some ways what mathematicians do has a lot in common with what artists achieve. A painter may know exactly what he or she is trying to paint, but getting the result is not so easy. A choreographer may know exactly what effect is needed, but fitting the steps to the music is what it is all about. In fact, let me compare an abstract group to a dance. The group can be represented in many ways, and the dance by many different dancers. Mathematicians frequently omit the distinction between an abstract group and some favourite way of representing it, but as W. B. Yeats wrote at the end of his poem 'Schooldays':

> O body swayed to music, O brightening glance,
> How can we know the dancer from the dance?

Treating groups in an entirely abstract way is done in serious mathematics books, but we don't need it. We need only think of a group of operations, such as permutations. Quite what the operations do does not matter – they can be represented in different ways according to the object they are acting on, and different objects can help in understanding the same group. Another good analogy is music. A piece of music can be accompanied by words, movement, or dance, or can simply be appreciated on its own. It is the same with groups. They can be seen as groups of symmetries, permutations, or motions, or can simply be studied and admired in their own right.

Sometimes an abstract group appears in two surprisingly different ways, and group theorists find this intriguing. We have already met a fine example: the symmetry atom of size 60 appeared as the group of all even permutations of five objects, but also as the rotation group of the dodecahedron. The connection between these two is something of an anomaly, but such anomalies yield larger anomalies, producing things such as the Monster that do not fit into any expected pattern. We shall find out more about this later.

In the meantime the study of group theory moved in an unexpected direction that later led to a 'periodic table' of symmetry atoms. The man responsible for the new development was a Norwegian pastor's son named Sophus Lie (pronounced Lee). Lie theory, like Galois theory, has since become a vital part of modern mathematics.

5

Sophus Lie

As the sun eclipses the stars by its brilliancy, so the man of knowledge will eclipse the fame of others in assemblies of the people if he proposes algebraic problems, and still more if he solves them.

Brahmagupta (598–670)

The most famous scientists have bold new ideas that take their subjects in directions hitherto undreamed of, and mathematicians are no exception. Galois was one of these, and another was Sophus Lie, whose work moved group theory on to radically new ground. Aiming to develop a Galois theory for differential equations – more on that later – he created groups in which one operation could be gradually transformed into another. These groups are infinite in size, but had a huge impact on the discovery of all finite symmetry atoms.

Sophus Lie had a personality and physique to match the strength of his work. One recent biographer has referred to him as 'the embodiment of an archetypical character in a theatrical drama – with his forceful beard, his sparkling green-blue eyes magnified by the stout lenses of his spectacles . . . a primal force, a titan replete with the lust for life, with audacious goals and an indomitable will'.*

Lie was born in 1842 in Oslo (known as Christiania at that time), but his family moved to the coast when his father was appointed pastor of a small town. His mother died a year later, when he was ten, and at 15 Lie left home to go to school in Oslo, and later to university there. As a student he was a great gymnast who loved jumping the wooden horse in the gym, and would also leap over real horses, placing his hands on the horse's back and springing over to the other side. Lie was also a great walker and went on long hiking treks. Fifty miles in a day was normal for him, and if he wanted to visit his family he simply walked the 36 miles to get there. One day he even walked the round trip just to get a book he needed.

At school and university, Lie specialized in science, but he found the last year of his studies rather tedious, and when his courses were finished in December 1865 he went home with no idea what to do. In March he wrote to a close friend, saying, 'When I bid you farewell before Christmas I believed that it was for now and all eternity, for my intention was to become a suicide. But I do not have the strength for it.'* Lie suffered from bouts of depression, but could also be extremely active and enthusiastic. That summer he went to a town further south to stay with his eldest sister and her husband, a doctor with a thriving medical practice. Sophus Lie was well known in the town for his various escapades, and that summer he decided to organize a swimming school for the doctor's son and the child's friends. They rowed the doctor's boat out into the fjord, with Lie in the stern ready to throw cold water on any boy who broke the unison of the stroke. To expedite the swimming lessons, Uncle Sophus strapped a lifebelt on his nephew and tossed him overboard. There was a fresh breeze coming down the fjord and it swept the boy off along the wave tops, where they lost sight of him. Fortunately the onlookers, watchful of the unpredictable Sophus Lie, followed the boy, jumped into another boat to pull him out, and wrapped the shivering little fellow in a coat. When Lie's boat pulled alongside, he demanded the child back, but they demanded his clothes to be handed over to their boat. As a recent biographer put it, 'Sophus Lie was said to have cursed them up and down, taunted them and threatened to smash in their skulls if the boy did not come over to his boat.'* However, the clothes were handed over and the whole town came to meet them when they got to shore. It was said that local mothers would later use Lie as a threat to keep naughty children in line: 'Now behave, or Sophus Lie will come and get you!'

At this time Lie still had no idea what to do with his life. He loved teaching, and had worked as a private tutor when he was a student, but he didn't want to become a school teacher. He worked as a teaching assistant in astronomy, his favourite subject at the time, but failed to get a permanent position – the astronomy professor couldn't stand his behaviour and was said to be outraged that on cold days Lie leaped over the apparatus to keep warm. He was once locked into the observatory – no one knows whether by accident or not – and simply jumped out of an upstairs window. Although he eventually lost the position, he continued his love for astronomy and gave a series of popular lectures on it.

Slowly, Lie got more interested in mathematics, and in the summer of 1868 there was a big meeting in Oslo. He attended, and heard of recent work by French, German, English, and Italian mathematicians. Thus far he had done no research work himself, and wrote later that 'To get involved in original scientific work never occurred to me. Above all, I was thinking of improving our mathematical pedagogy. I was preoccupied considerably with this.'*

That autumn he started his own research project in geometry, and published a paper at his own expense. It was translated into German and accepted by an important journal founded by Leopold Crelle in 1826 and edited in Berlin. This won him a grant, and the next autumn he went to Berlin. There he met a young German mathematician named Felix Klein, who later became one of his most influential supporters. Lie and Klein formed a very productive relationship because Klein loved hearing of new things to fit into his big picture of mathematics, while Lie loved to pursue his own idiosyncratic ideas. The result was that Lie explained his ideas to Klein, Klein reacted to them, Lie responded to Klein's reactions, and a very useful discussion developed.

In early 1870 Lie went to Paris, and Klein joined him later, at the same hotel. In Paris they met, among others, Jordan, whose magnificent treatise on groups of permutations was just rolling off the presses. The visit was very stimulating, but ended suddenly in mid-July when war with Prussia broke out. Lie and Klein soon left, which was a good thing because in September Prussian forces surrounded Paris, and kept it under siege. The minister of war left the city by balloon to join a government in exile at Tours, but French troops were unable to lift the siege, and the inhabitants of Paris were in some trouble. As winter set in, animals from the zoo were auctioned off for meat.

Klein had returned immediately to Germany, but Lie decided to visit Italy to see a mathematician named Luigi Cremona. He intended to hike there, but had gone no further than Fontainebleau when he was arrested and imprisoned on suspicion of being a German spy. There are several stories about this. Apparently he had a good technique for keeping his clothes dry in the rain by removing them, putting them in his backpack and carrying on. This may have attracted some attention, but so did the fact that he sang Norwegian songs – which would have sounded German to some people – and carried a sketch book that he used to draw interesting features in the landscape. His mathematical papers were taken to be coded messages, 'lines' and 'spheres' being interpreted as 'infantry' and 'artillery', and he was in prison for a month before a mathematical friend named Gaston Darboux came with a letter from the interior ministry to have him released.

When Lie finally got back to Norway in December, he was well known as the scientist who hiked through France and was imprisoned as a suspected German spy. He had fun telling the story, but soon got down to serious work, and by next summer completed an excellent doctoral thesis, and applied for a position that had recently opened up in Sweden. This prompted the Norwegian Parliament to debate establishing a new chair for him. When the matter came up for discussion he tried to push his way forward in the public gallery and was thrown out, but all went well. It was a resounding vote of confidence in his abilities, and Lie quickly became a big name in mathematics.

Lie admired Galois's work enormously and wanted to do for differential equations what Galois had done for algebraic equations. Differential equations involve rates of change, and have wide uses in economics, engineering, physics, and other areas. They are very different from the algebraic equations that Galois studied, where each equation has a finite number of solutions. By contrast a differential equation has infinitely many solutions. For example,

Sophus Lie in 1886 shortly before he left Norway to take up the chair in Leipzig.

a single differential equation describes a vibrating string, but its solutions depend on where the string was stopped, and this can be continuously varied, giving infinitely many solutions. Like Galois, Lie wanted to consider all the solutions together – and then see how one morphed into another as the initial parameters changed.

This led him to groups of 'continuous transformations', where one operation could gradually transform into another. Gradual changes pass through infinitely many stages – as you accelerate your car its speed never jumps suddenly, so it passes through infinitely many tiny changes. This makes Lie's groups of continuous transformations infinite in size, and it is remarkable that they are so useful in finding the finite atoms of symmetry.

Why this should be the case is an interesting question for philosophers. It is similar to a question about physics. If the universe is made of tiny quantum particles that can only jump minutely from one state to another, why do we use continuous mathematics, such as string theory, to describe it? In Chapter 7 we shall find a way back to finiteness – in mathematics, though not physics.

Lie's work on differential equations used multidimensional geometry. This sounds hard, and those trained in the traditional methods of differential equations found it difficult to follow him. Initial incomprehension like this occurred with Galois's work, but unlike Galois, Lie lived to promote his methods and encourage students.

His use of geometry came about by treating the parameters of an equation as coordinates. The idea of doing geometry by coordinates had now been around for over two hundred years, after being pioneered by René Descartes, the famous French philosopher who lived in the first half of the seventeenth century. In 1637 Descartes produced his famous *Discourse on Method*, which he wrote in French rather than Latin. He wanted people without a scholarly training to be able to follow his arguments, believing that everyone could tell the true from the false by natural reasoning. His *Discourse* had an section on geometry that could be understood independently of the rest, and introduced coordinates (the term Cartesian coordinates is derived from the name Descartes).

Here is how it works in two dimensions. Each point of the plane is specified by two numbers, a and b, measuring distances along the two axes.

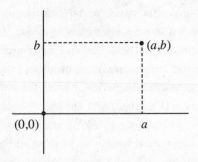

In three dimensions each point has three coordinates (for, example two horizontal and one vertical), so we can think of three dimensional space as the set of triples (a,b,c) where a, b, and c range over all numbers. It is a small step from here to create four dimensions: simply take the points to be all quadruples (a,b,c,d). Once you have the hang of this you can create five, six, seven, or more dimensions.

This sounds deceptively simple, but I assure you that mathematicians do not easily see pictures in four or more dimensions. Like sculptors, most are fairly good at three dimensions, but four dimensions is another matter entirely. We might get quite adept at it, with practice, but increase the number of dimensions and pictures become impossible. What we *can* do, however, is to work by analogy. We know what it means to go from a two-dimensional plane to three-dimensional space, and we can think in a rather formalized way of adding a new dimension, and then another dimension beyond that, and so on. For example, in a plane, two lines that are not parallel must intersect, but in three dimensions this is not necessarily the case. Going from two dimensions to three opens new possibilities, and it is similar when you can go up from three dimensions to four. In three dimensions a line and a plane intersect unless the line is parallel to a line in the plane. But in four dimensions a line can be parallel to nothing in a plane, yet they may not intersect. It is as if the line can cross the plane without touching it, but be careful of thinking in those terms. In four dimensions a plane does not have two sides, any more than a line has two sides in three dimensions.

Before we ask whether geometry in four or more dimensions has anything to do with reality, it is worth asking a similar question about our usual Euclidean geometry in two or three dimensions. This is a geometry in which points have no size and lines have no thickness. It is a convenient abstraction, but it *is* only an abstraction and is not the real geometry of the universe, whatever that might be. For example, in the real universe elementary particles have a certain finite size, however small, and lines and planes have a certain thickness. Quantum theory asserts that we cannot go down beyond a certain finite quantum limit; there is a graininess to the universe that we cannot seem to beat. The abstract world of Euclid's geometry has no graininess in this sense, but it is extremely useful, and we don't dismiss it just because it isn't physics. Nor should we dismiss space of many dimensions just because it isn't what we seem to perceive.

It is very useful in practical applications of mathematics because each variable in a problem has a certain degree of freedom, and each new degree of freedom adds a new dimension. If you have two or three variables the problem can be described in geometry of two or three dimensions, but with more variables you need geometry in more than three dimensions. In Lie's case an equation could have many parameters, so he needed multidimensional space. It is a vital component in modern mathematics and we shall meet it again later.

After Lie assumed his new chair at the university, he pursued his research programme with vigour. In 1872 he visited Klein, who was now in Erlangen, and when he returned that Christmas he got engaged to a young lady of 18, and they were married in 1874. His work progressed well, and he was soon led to the concept of 'finite, continuous groups' – now called Lie groups. The word continuous meant that each transformation in the group could be continuously modified, and the word finite meant there were finitely many degrees of freedom for these modifications. To put it another way, there were finitely many coordinates, and Lie's groups are finite dimensional.

Here is an example. Take a circular disc with its centre fixed. The disc can rotate around its centre, and one rotation can be gradually transformed into another. The rotations form a Lie group, which we can interpret geometrically as a circle, each point on the circle representing a rotation: one point represents no rotation, and as you move round the circle the angle of rotation increases, until you are back where you started. A circle is a line that curves round on itself, and this is an example of a one-dimensional Lie group.

For a higher dimensional example, think of an object moving in a flat plane. As we slide the object around there are two degrees of freedom for the motion because the plane is two-dimensional, and if we allow the object to rotate this gives a third degree of freedom. The group of all motions in the plane, sliding and rotating, is three-dimensional: two flat dimensions for the plane itself, and one curved dimension for the rotation. Geometrically one can view this group of motions as a three-dimensional curved space, which is impossible to imagine – unless you have had lots of practice. This is for motion in two dimensions, but if we replace the two-dimensional plane by three-dimensional space it becomes even more complicated. A tennis ball moving and spinning in three dimensions has six degrees of freedom,* so describing its motion uses six dimensions!

Now I should emphasize that while Lie and Klein were using geometry, they did not take it as their starting point. As Klein wrote in a letter to another mathematician in 1870, 'we do not think of the geometrical configuration as given, and ask about the transformations; rather we consider the system of transformations as given, and ask about the geometrical configurations'.* This was analogous to Galois's attitude; he used groups of permutations, but that was not his starting point. He started with equations. And so did Lie, differential equations in his case, but his geometric insight then led him to groups of continuous transformations.

In the meantime a German high school teacher, Wilhelm Killing, who was five years younger, was pursuing similar ideas on groups of transformations. He wrote a long essay in 1884, sent it to Felix Klein (who now occupied the chair of geometry in Leipzig), and received an immediate response informing him of Lie's work. Killing then wrote to Lie in Norway, requesting copies of his papers, but got no response. He had no access to papers in Norwegian journals, so Killing wrote again to Klein, who wrote to Lie, and the next year papers were sent, but only on loan, and when Killing asked to keep them longer he got no response. Being a very sincere man, he felt he had to return them before fully digesting the results.

As for Lie, he felt a bit cut off in Norway, and in September 1884 Klein despatched a young German mathematician to Norway to help him out. The young German from Leipzig was named Friedrich Engel (not to be confused with Friedrich Engels, who worked with Karl Marx). Engel returned home next summer with a huge manuscript. The following year, 1886, Klein moved from Leipzig to a chair in Göttingen – where he built up a magnificent school of mathematics – and Lie was persuaded to leave his beloved Norway and take Klein's old chair at Leipzig. It seemed an excellent move for Lie, who could now work with Engel on the massive book they were writing together. Students could easily come from other parts of Germany to work with Lie, and since he was well known in Paris, and was very much at home with their mathematical attitudes, they sent some of their best students to Leipzig. Everything seemed to be going swimmingly.

Meanwhile, Killing got into communication with Engel, who, as soon as he read the long 1884 essay, wrote back to assure Killing that 'you have also discovered transformation groups in Lie's sense'. Killing responded two days later to urge Lie and Engel to get on and publish their results, '[This] will, I hope, induce you and Lie to publish more rapidly. Naturally I do not want to enter into competition over this theory, but . . . I have been led to results that at least until now have not been published.'

Killing soon went to visit Lie in Leipzig, but it seems the two of them did not hit it off. Lie's reaction is recorded in a letter to Klein: 'Killing was just here. He has some really nice ideas. In many other respects, however, he does not make a solid impression.' Their relationship did not improve, and animosity developed later as Lie's mental health broke down, but let us first see where Killing came from and what he achieved.

He began university studies in 1865 in north-western Germany at a university (in Münster) that had no mathematicians; the subject was taught by observational astronomers. His fellow students were also a disappointment, and Killing recalled that they 'showed almost no interest whatsoever in science itself; they wished (with very few exceptions) to study only what was needed for the examinations'. After four semesters he abandoned Münster and went to Berlin, which was the centre of mathematics in Germany at the time. Killing appears to have been a good-hearted man because he interrupted his studies one year to teach, sometimes as much as 36 hours a week, at a school in a small town where his father was mayor. The school had been threatened with closure and Killing taught all subjects. He then resumed his studies in Berlin, got his doctorate, and went back into teaching. In 1880 he became professor at an academic training establishment for Catholic clergymen.

From this unpromising position, Killing wrote his 1884 essay, and then started pursuing the idea of classifying Lie's groups of continuous transformations – what we now call Lie groups. This means finding all the 'simple' ones and placing them in families. In rapid succession he wrote a series of three papers on this classification, and sent them to Klein for publication – Klein had become an editor of *Crelle's Journal*, where Lie's first paper had appeared. It is a fine publication that exists to this day. The third paper was sent in October 1888 and published the next year.

Killing had discovered a 'periodic table' of Lie groups. He placed them in seven families: *A* to *G*.[1]

$A1$	$A2$	$A3$	$A4$	$A5$	$A6$	$A7$	$A8$	$A9 \ldots$
$B1$	$B2$	$B3$	$B4$	$B5$	$B6$	$B7$	$B8$	$B9 \ldots$
	$C2$	$C3$	$C4$	$C5$	$C6$	$C7$	$C8$	$C9 \ldots$
			$D4$	$D5$	$D6$	$D7$	$D8$	$D9 \ldots$
	$G2$		$F4$		$E6$	$E7$	$E8$	

The number in each case is called the rank of the group, and is related to the number of dimensions it can operate in. The higher the rank, the higher the number of dimensions, so as we move from left to right in this table the number of dimensions increases. The *A* family is the simplest; families *B, C*, and *D* are more complicated but all relatively similar to one another. These four families, *A, B, C*, and *D*, taken together, are called classical. The five exceptions of types *E, F*, and *G* terminate at rank 8. There are no Lie groups of types *E*9, *F*5, or *G*3; if one tries to create such things they become infinite-dimensional. On the other hand, decreasing

---

[1] The reason I have not listed *C*1, *D*1, *D*2, or *D*3 is that either they are not 'simple' or they are included in those already listed. For example, *D*3 is the same as *A*3.

the rank only yields entries already in the table – for instance, $E5$ is the same as $D5$.

Killing had worked very swiftly and he knew only too well that some of the theoretical analysis was inadequate. He wrote to Klein that 'If I said I was satisfied with the stated results I would be lying. I have attempted on many occasions to find an error in the proof, but so far without success. . . . I thought it best to publish the results with the proof as quickly as possible, since only then can a serious thorough examination take place; and it is of the greatest importance to me that complete certainty about these questions be achieved as quickly as possible.'

The importance of Killing's work was immediately clear to Lie, who wrote that it 'contains results of the greatest importance, if only everything is correct'. Later he wrote to Klein that 'Killing has done beautiful research. If, as I believe, the results are correct, he has performed an outstanding service. Generally speaking, now the theory of transformation groups . . . will reign over vast areas of mathematics.'

It is really unfortunate that Killing, working as he did at a school for the training of future clerics, had no research students. He had no junior mathematicians to clean up after him, and tie up some of the loose ends, and he regretted it, 'If only I had students in mathematics, I would allow many excursions into structure. The groups of ranks 4 to 8 provide easy material for seminar works.'

Lie, on the other hand, had plenty of students in his new position at Leipzig, but he found the teaching to be a very different matter from his easy experiences in Norway. As he explained to a Norwegian friend, 'While in Norway I hardly spent five minutes a day on preparing the lectures; in Germany I have to spend an average of three hours. The language is always a problem and above all, the competition implies that I have to give eight to ten lectures a week.' In 1888 the first volume of his massive book with Engel was published, and in 1889 the second volume went to press, the same year that Killing published the third of his classification papers. Then, in late 1889, Lie had a nervous breakdown, and was admitted to a mental hospital, where he remained for seven months.

In the meantime, Killing's discomfort with his proofs was well founded. The results were correct, but an error in the first paper contaminated the other two papers, and a new approach was needed. Killing left the matter to others and returned to his main love in mathematics – the foundations of geometry. That was what had driven him to study groups of continuous transformations in the first place, and he was now offered the chair of mathematics in Münster, the university where he had first been a student.

He was now free to pursue his original research programme, but it is an extraordinary thing in life that creative people often do their greatest work when circumstances are at their most difficult, and this seems to be the case with Killing. In his new comfortable position at Münster, he published a book on the foundations of geometry, but as Engel wrote, 'Killing's latest opus on the foundations of geometry contains real nonsense'. Nevertheless, Engel strongly supported Killing for a book prize because he knew how important this work was for its author, and how it had led him to his really great work, the classification of Lie groups.

As Killing fades out of the picture we are left with two problems. His results were correct, but the theoretical analysis had to be fixed up so that their truth could be readily verified. The other problem was to construct all the groups in the families Killing had classified, to verify that they all exist.

It remained to Élie Cartan (1869–1951), a young graduate student in Paris, to put these things right. Cartan was the son of a blacksmith plucked from obscurity by an inspector of schools. This gentleman encouraged one of the teachers to give the boy special coaching and he won a full scholarship to a fine boarding school. From there he went from strength to strength and in 1888 entered the École Normale (the same place Galois had been a student, and now the premier establishment in France for mathematics). In 1892 Cartan returned to Paris from a year's military service, and took rooms with another student who had just returned from studying in Leipzig. This student told Cartan about Lie's groups and Killing's classification. Cartan became fascinated and decided to devote his doctoral thesis to this topic.

As for Lie, in autumn 1892 he had fully recovered from his long depression, and was able to take delight in the fact that his theory was now all the rage in Paris. As one senior French mathematician, Émile Picard, wrote to him, 'You have created a theory of major importance that will be counted among the most remarkable mathematical works of the second half of this century.' By 1893 Picard could write again that 'Paris is becoming a centre for groups; it is all fermenting in young minds, and one will have an excellent wine after the liquors have settled a bit.'

In 1893 the final volume of his great book with Engel was published, and Lie himself visited Paris. Cartan was thrilled to meet the great man and wrote later, 'I can never forget all that I owe to the great Norwegian scientist whom I had the honour of seeing often in Paris in 1893.' Cartan completed his doctoral thesis the next summer, filling the gaps in Killing's work, and confirming the periodic table that Killing had found.

Cartan was just the man for the job. He had a great talent for abstract structural reasoning and this helped him to clarify and develop Killing's ideas. Some of the technical details were renamed and new details were added, and the result is now known as the Killing–Cartan classification.

Abstraction is a vital part of mathematics. It is essential for simplifying and merging difficult technical ideas, so that new progress can be made. Cartan's attitude to abstraction is well illustrated by a talk he gave much later in life (1940 in Belgrade):

> More than any other science, mathematics develops through a sequence of consecutive abstractions. A desire to avoid mistakes forces mathematicians to find and isolate the essence of problems and the entities considered. Carried to an extreme, this procedure justifies the well-known joke that a mathematician is a scientist who knows neither what he is talking about nor whether whatever he is talking about exists or not.

In 1894, the year Cartan completed the classification, the Norwegian National Assembly established a chair for Lie. This was very welcome because he wanted to return to his homeland. But his wife and daughters had friends in Leipzig and were reluctant to leave, so Lie stayed on, until by 1898 it became clear that he was now suffering from pernicious anaemia, and it was high time to go. The family returned to Norway. Lie delivered some lectures in the autumn of that year but soon had to give up, and was reduced to holding seminars in his home. They too soon had to stop, and Lie died in February 1899.

6

Lie Groups and Physics

How can it be that mathematics, being after all a product of
human thought independent of experience, is so admirably
adapted to the objects of reality?

Albert Einstein

While Lie was engaged on his research, the structure of classical
physics still seemed fairly secure, but this did not last. When Lie
died, shortly before the end of the nineteenth century, the edifice
of classical physics was starting to crack. New observations at
the microscopic scale, within atoms, and at the cosmic scale eventually led to the development of quantum theory and general
relativity, and Lie's work found a ready audience among some
young physicists, as we shall see.

The originality of his ideas had opened up a new field of mathematics, and in 1922 at a lecture to the Norwegian Mathematical
Association, his erstwhile collaborator Engel expressed himself
enthusiastically on this point:

If the power of invention is the true yardstick of mathematical
greatness, then Sophus Lie must be reckoned as the first among all
the mathematicians of the time. The new fields that he has opened
for mathematical research are so extensive, the methods he has

created, so fruitful and far-reaching, that only extremely few can, in this respect, stand to be measured with him.*

Since Lie was such a towering figure, his name has become associated with a broad swath of modern mathematics, all under the heading of Lie theory. This means not just the Lie groups themselves, but algebraic structures called Lie algebras, along with other material related to the classification work of Killing and Cartan. As the twentieth century progressed, Lie theory seemed only to increase in importance, and in 1974 the French mathematician Jean Dieudonné wrote, 'Lie theory is in the process of becoming the most important field in modern mathematics. It had gradually become apparent that the most unexpected theories, from arithmetic to quantum physics, all circle around this field, as around a giant axis.'* Quantum physics has made extensive use of Lie theory, and I will mention some applications in this chapter. Later in the book it will connect up with the Monster, via string theory, which is a way of combining quantum physics and general relativity theory.

The theory of relativity, for which Einstein is so famous, began to emerge in the late nineteenth century from the study of electricity and magnetism. These were seen as manifestations of something called electromagnetism, which propagates as a wave: radio waves, X-rays, and light being examples. Experiments had shown that electromagnetic waves travel at a speed – the speed of light – that appears the same to all observers no matter how fast one is travelling relative to another. This apparent paradox led to the *special* theory of relativity, where three dimensions of space and one dimension of time were replaced by four-dimensional space-time. The geometry of space-time was studied mathematically by Hermann Minkowski, in a way that we shall meet again in Chapter 17.

The idea that all motion was relative begged the question of acceleration. Was that to be taken as relative to the observer too, or was there a state of no acceleration that could be agreed by all observers? As everyone knows, if you accelerate quickly in a car or a plane, you are pulled back into your seat. The force is real, so presumably the acceleration is real too, but what do you measure it against? Suppose you are ensconced in a spaceship in the deep reaches of space, and the spaceship accelerates. You will feel a pull. Is that any different from the pull of gravity? As Einstein saw it, there was no difference. The two were indistinguishable, and to make sense of this he had to curve space-time. Minkowski's geometry, which originally had no curvature, now had to bend under the gravitational effect of massive objects, rather as a heavy person sags a mattress.

Special relativity, along with the curvature of space-time from massive objects, was called *general* relativity, and was soon used to explain an anomaly in Mercury's orbit round the sun. General relativity became an accepted part of physics, but Einstein wanted to see gravitation and electromagnetism understood on the same basic principles, and as one recent biographer of Lie has written, 'Élie Cartan was one of his most important partners in this discussion. In a three-month period, from December 1929 until February 1930, they exchanged twenty-six letters. Einstein was seeking mathematical expertise from Élie Cartan: not the least fascinating aspects were Cartan's theoretical interpretations of general presentations of space based on representations of Lie groups.'* In this context a representation of a group refers to a way it can operate in multidimensional spaces, and this became vital in quantum theory, and for understanding the way electrons are arranged in atoms.

Electrons, like other quantum particles, have the strange property that they can behave as both particles and waves. First think in terms of particles. An electron has a negative electric charge, and in an atom the electrons orbit a very small, positively charged nucleus. The Danish physicist Niels Bohr created this model of the atom to fit in with the experimental evidence, and it was an excellent model, except for one thing. Classical physics foresaw that an electron travelling along a curved path must emit radiation. This would decrease its energy and it would end up spiralling in to the nucleus – according to classical physics, atoms couldn't exist. This problem was overcome by supposing that energy could not gradually dissipate, but could only be emitted in multiples of a small quantum of energy. An electron could not gradually change its energy or its orbital pattern. There would always be a small quantum jump. This would stop an electron from spiralling in to the nucleus; it would reach an orbit of lowest energy where it could go no lower.

A quantum jump breaks the idea of continuity, suggesting that Lie's groups would have no useful role to play, but on the contrary, they are very useful in quantum theory because elementary particles also exhibit themselves as waves. An electron appears as a wave, smeared out over space, but it is also a particle: if you manage to grab it, you grab all of it. You can never get only part of an electron. Quantum theory is a mysterious business, and as Niels Bohr himself said, 'Anyone who is not shocked by quantum theory has not understood it.' With the passing of time one might expect that the subject would become easier to comprehend, but as the late Richard Feynman said in a comment in 1965, 'I think I can safely say that nobody understands quantum mechanics.'*

The wave nature of an electron means that as it orbits the nucleus of an atom, it is not at all like a planet orbiting the sun. It has to be treated as a wave encircling the nucleus, and that brings Lie groups into the picture. An atom exhibits spherical symmetry, and this suggests that the Lie group of rotations in three-dimensional space should be important in the structure of electron orbits. In the simplest case an orbit has spherical symmetry, the electron being smeared out around the nucleus like the surface of a balloon around its centre.

Now a very important principle must be taken into account. Two electrons in the universe cannot be in the same state. Within an atom this means that no two electrons can have the same energy and the same orbit – unless they have opposite spins. An electron spins, either one way or the other, either spin-up, as they say, or spin-down. When an orbit is spherically symmetric two electrons are allowed, one spin-up and one spin-down.

The group of rotations leaves a spherically symmetric orbit unchanged, and the operation of the group is said to be one-dimensional. But most electron orbits in a large atom are not spherically symmetric, and the group of rotations can change one into another. In this case the operation of the group is more than one-dimensional – there is more than one degree of freedom. The number of degrees of freedom – or mathematically speaking the number of dimensions – has to be an odd number: 1, 3, 5, 7, etc. This is a mathematical fact about the Lie group of rotations in three dimensions.

For each degree of freedom there can be just two electrons, one spin-up and one spin-down. When there are three degrees of freedom this means that six electrons are allowed, and these electrons form what is called an electron orbital. Since the number of degrees of freedom is an odd number, the size of each electron orbital is twice an odd number: 2, 6, 10, 14, etc. This means, for instance, that the existence of electron orbitals filled by ten electrons occurs for a good mathematical reason – it is because the group of rotations in three dimensions has an operation in five-dimensional space!

These numbers for the size of electron orbitals are critical in creating the periodic table of elements. In other words, mathematical properties of the group of rotations in three dimensions are determining factors in the structure of atoms.

This is one use of Lie groups, but there are other Lie groups also involved in quantum phenomena. Physicists believe there are four fundamental forces of nature: gravitation, electromagnetism, the weak nuclear force, and the strong nuclear force. The first one curves space-time, as described by Einstein's general theory of relativity. The other three are quantum forces, and to each one physicists associate a Lie group, which they call a gauge group in this context.

The gauge group for electromagnetism has one degree of freedom, and this corresponds to the fact that one quantum particle – the photon – acts as a medium for the force.* The weak nuclear force, which is responsible for holding a neutron steady, has a gauge group with three degrees of freedom, corresponding to the three particles that mediate the force.* The strong nuclear force, which is responsible for holding together the

76

nucleus of an atom, has a gauge group with eight degrees of freedom, corresponding to eight different gluons that mediate the force.*

One of the things physicists want to understand is the relationship between the three quantum forces: strong nuclear, weak nuclear, and electromagnetism. Gravity is a separate problem because there is no theory of quantum gravity yet. By seeing all three as manifestations of a single force that perhaps existed at the birth of our universe, physicists hope to get a deeper understanding of quantum phenomena. If they knew exactly how to fit their Lie groups together into one larger Lie group – there are many ways of doing this, but the right one has to hold up under experimental evidence – then they might discover a deeper symmetry between all the elementary particles.

When quantum theory first hit the research journals in 1925, the new advances were largely taking place in Germany, which was a powerhouse of mathematics and physics at the time, but the conditions for creating advanced new ideas in science didn't last because in 1933 Hitler came to power. The Nazi government soon destroyed intellectual life, and when a Nazi minister visited Göttingen in the 1930s, and asked David Hilbert – a famous mathematician with a chair at the university there – how mathematics was doing now they had rid the place of the Jewish influence, Hilbert's response was 'There is no longer any mathematics in Göttingen.' Indeed, it was the end of first-rate mathematics in Germany for a long time. The Institute for Advanced Study had just been founded in Princeton, and several of the best minds went there: Albert Einstein and Hermann Weyl, for example, had both spent time in Göttingen. Weyl was a great proponent of Lie's groups and their use in physics.

The centre of gravity in mathematics was shifting, but even before this shift occurred, an important new development in our story had occurred in America.

7

Going Finite

The infinite we shall do right away. The finite may take a
little longer.

Stanislav Ulam (1909–1984)

Lie's groups of transformations, in Chapters 5 and 6, form the
prototypes for most of the finite symmetry atoms. On a philo-
sophical level this is surprising because Lie's groups embody a
continuity that makes them infinite in size – one transformation
can gradually morph into another, just as we can gradually morph
one number into another by increasing or decreasing it. But finite
versions were created by a young American mathematician named
Leonard Eugene Dickson.

Dickson was the first graduate student in mathematics at the
University of Chicago, which is now one of the top academic
establishments in the world. In 1896, with a fresh PhD in his
pocket, he went to Europe to learn more, visiting Paris and
Leipzig. In Paris he met the young Élie Cartan, he of the Killing–
Cartan classification for Lie groups; and in Leipzig he met
Lie and Engel, soon after their great three-volume treatise had
been published. Paris and Leipzig were *the* places to learn about
these things, and when Dickson returned to America, he set about
creating finite analogues for most of the Lie groups in the periodic

table, providing, in the event, a huge collection of symmetry atoms.

To achieve this, Dickson replaced the usual system of numbers by a finite system. This finite system of numbers had to be self-contained in the sense that the sum, difference, product and quotient of any two numbers must be another number in the same system. In other words you could do all the usual operations of arithmetic and stay within the system. Let's see how this is possible.

In any sort of arithmetic, finite or not, the only numbers we insist on having are 0 and 1. If our system allows addition we need $1 + 1$ and $1 + 1 + 1$ and $1 + 1 + 1 + 1$, and so on. This seems to demand infinitely many numbers, so how can a finite system be self-contained? It seems impossible, yet we already know one: the twelve digits on a clock face. You add one digit to another by moving in a clockwise direction, and you subtract one from another by moving counter-clockwise. For example, on a clock face 5 plus 9 is 2: if you add five hours to 9 o'clock you reach 2 o'clock. As an example of subtraction, if you move five hours back from 9 o'clock you reach 4 o'clock, so $9 - 5 = 4$. This looks like ordinary arithmetic because in moving back five hours from 9 o'clock we didn't go past the 12 o'clock point. When we do, for example, by counting back nine hours from 5 o'clock and reaching 8 o'clock, we get the equation $5 - 9 = 8$. This looks strange at first, but the point is that the twelve numbers on a clock face form a system that is self-contained in terms of addition and subtraction. I shall call it 12-cyclic arithmetic.

Of course there is nothing exceptional about 12; any whole number will do. Let us try the same thing with 7, and create

7-cyclic arithmetic. Place seven numbers – 0, 1, 2, 3, 4, 5 and 6 – in a circle.

These numbers can be added to one another, or subtracted from one another, by moving around the circle: clockwise for addition, and counter-clockwise for subtraction. For example, in 7-cyclic arithmetic $4 + 5 = 2$ because if you move clockwise five steps beyond 4 (or four steps beyond 5), you reach 2. As an example of subtraction notice that $4 - 5 = 6$, because if you move counter-clockwise five steps back from 4 you reach 6.

I am using 7 as an example, rather than 12, because 7 is a prime number. It has no divisors except 1 and itself. This makes a big difference when it comes to multiplication. For example, on the usual clock face 12 is identical to 0, so 3×4 is 0. The product of two non-zero numbers is zero. This leads to problems, and to avoid them mathematicians tend to do cyclic arithmetic only with prime numbers such as 7. For instance, in 7-cyclic arithmetic $3 \times 4 = 5$, because 12 is the same as 5. This looks odd if you have never seen it before, and it takes some getting used to.

Division looks strange too, but first we should understand what we mean by division. As a child I was at first baffled when faced with questions such as $6 \div 3 = \ldots$ until my teacher explained that I simply needed to find a number that, when multiplied by 3, would give 6. I soon found that division problems could be solved by using a multiplication table and working backwards, but I felt a bit let down. Division was supposed to be difficult, and looking up the answer seemed rather too easy.

Now try division in 7-cyclic arithmetic. What is $6 \div 3$? Why, it's 2 of course. Now try $5 \div 3$. Is this possible? If so, it has to be a number which, when multiplied by 3 gives 5. This may seem impossible; but the answer is 4, because we noticed earlier that $3 \times 4 = 5$ in 7-cyclic arithmetic. This implies that $5 \div 3 = 4$, and $5 \div 4 = 3$. Of course there should be a way of doing division without knowing the answer in advance, and there is. The method shows, by the way, that division is always possible in p-cyclic arithmetic whenever p is a prime number (of course you can't divide by zero; that is not an option in any system of numbers).

Finite arithmetic enables mathematicians to go from continuous things that rely on the real numbers, to finite things. It is as basic to mathematics as the telescope is to astronomy, and what Dickson did was to telescope some of Lie's groups down to finite versions. He dealt with families A, B, C, and D because they could all be treated as symmetry groups in Euclidean space, and Dickson could replicate this symmetry as he went down the telescope from the real numbers to p-cyclic arithmetic (p being a prime number). For example, in the Lie family of type A (Chapter 5) there are symmetry groups denoted $A1$, $A2$, $A3$, etc. Each of these can now appear in different finite versions, one for each prime number: 2, 3, 5, 7, 11, etc. In other words, there is an $A1$ group for each prime number, an $A2$ group for each prime number, and so on.*

Some of this work was not new. For example, Jordan, in his treatise of 1870, had already used cyclic arithmetic with the A-family, and Galois himself had done it with $A1$. He had also extended cyclic arithmetic to a more general form, often known as Galois arithmetic in his honour.

Dickson's symmetry atoms in families *A, B, C,* and *D* – the classical ones – cover most cases, but not all. He managed later to handle two non-classical cases, but a uniform method for *all* the Lie families took over half a century from the time of Dickson's book. We shall see why later. In the meantime I want to explain how geometry with a finite system of numbers plays a vital role in our modern electronic world.

The simplest finite system of numbers is 2-cyclic arithmetic. There are only two numbers: 0 and 1, and $1 + 1 = 0$. If you add 1 to itself an even number of times you get zero; an odd number of times and you get 1. With this system of two numbers you can do addition, subtraction, multiplication, and division (of course you can only divide by 1 – dividing by 0 is not allowed in any system). Having just two numbers 0 and 1 may seem a bit trivial, but 2-cyclic arithmetic is *extremely* useful, because computers operate using strings of zeros and ones.

Credit card numbers, bar codes at the supermarket, and many other sequences of digits, are read and processed electronically by first turning them into sequences of ones and zeros. When these numbers are read there is always a chance of making errors, so a certain amount of redundancy is built in. For example, the last digit on a bar code is a 'check digit'. If you alter it, or if one of the digits in the code is misread, then the bar code will almost certainly be invalid. There won't be any goods having that bar code. If you make a single error in giving your credit card number you will almost certainly have given a number that does not belong to any credit card.

This built-in redundancy means that we are not using all possible sequences of numbers. Those that we *are* using can be chosen in such a way that simple errors and misreads can be quickly corrected. Part of the trick in error-correction is to use geometry. Here's the idea. Suppose all the sequences we are using give points on a plane in three-dimensional space. If you read or receive a point that is slightly off the plane you know it's an error, and you correct it by moving to the closest point on the plane. This technique works really well with 2-cyclic arithmetic, but we need more than three dimensions. Here is why.

In our usual geometry of three dimensions each point has three coordinates, and each coordinate is a real number. When we replace the real numbers by 2-cyclic arithmetic, each coordinate is either 0 or 1. This means there are only two choices for each coordinate, so the number of points is just $2 \times 2 \times 2 = 8$. Eight points are not much. They are not remotely enough for everyday uses. My credit card, for instance, has a code number consisting of sixteen digits (each from 0 to 9). Converted into a sequence of ones and zeros, it needs 54 digits. A sequence like this can represent a point having 54 coordinates, in other words a point in 54-dimensions.

Higher dimensional spaces are extremely useful for practical applications of mathematics, and we shall meet them again on the road to the Monster. In the meantime I want to emphasize that group theory – the study of symmetry – was no longer confined to France, Germany, Norway, and the USA. Britain joined in, with enormous contributions from a man named William Burnside. He was born in 1852 and died in 1927 at the age of seventy-five.

Burnside was just four years old when his father died, and his family was in straitened circumstances. Being a very bright lad he won a scholarship to Christ's Hospital, a school that accepted only boys and girls from poor families. It was founded in the sixteenth century when Grey Friar's monastery in the City of London was turned into an educational establishment for London's many street children. The school acquired a mission to educate children from families that were poor or had fallen on hard times, and winning a scholarship to Christ's Hospital became a great honour.

William Burnside studied there until 1871, when he went on to Cambridge University. After completing his degree he spent the next ten years coaching in mathematics and rowing, but towards the end of this time he started publishing research papers, and at the age of thirty-three was appointed Professor at the Royal Naval College in Greenwich. Burnside didn't take an immediate interest in group theory, but when he did his contributions were wonderful, and in 1897 he published a book with the title *Theory of Groups of Finite Order*. It became a classic of mathematics, and in the preface he wrote:

> The subject is one which has hitherto attracted but little attention in this country; it will afford me much satisfaction if, by means of this book, I shall succeed in arousing interest among English mathematicians in a branch of pure mathematics which becomes the more fascinating the more it is studied.*

Burnside continued turning out excellent results, and in 1904 he published an important theorem about 'simple' groups – the atoms of symmetry. This theorem showed that if a 'simple' group is not prime cyclic, then its size must be divisible by at least three different prime numbers. For example, the smallest one has size

60, which is divisible by the prime numbers 2, 3, and 5. The next smallest has size 168, which is divisible by 2, 3, and 7. Burnside proved that this must be the case. If a group is not prime cyclic, and its size is divisible by only two different prime numbers, then it cannot be 'simple'; it must be composite.

This theorem of Burnside remains a famous result today, over a hundred years later, showing that mathematics offers a greater chance of immortality than almost any other subject. A theorem, once proved, is good for all time, and Burnside's theorem was a culmination of lesser results, which could then be safely forgotten. He and others had proved special cases of the result during the late nineteenth century, but the new theorem in 1904 overwhelmed this earlier work, and did so in a beautiful way. As one mathematician has written recently, 'Burnside's proof is very short and elegant – one of the great gems of group theory. [Other] proofs were found after a great deal of effort in the 1960s and early 1970s, but even the shortest of these cannot compare in elegance and comprehensibility with the original.'*

Elegance and clarity are markers of excellent mathematics, and Burnside had used a sophisticated new technique called 'character theory' that we shall hear more of later. Other proofs of his theorem were less elegant because their authors were trying to avoid 'character theory'. This is rather like going from Europe to China without using the modern convenience of air travel. You can do it, and it may be quite an interesting journey, but it will take much longer.

Mathematics makes progress by proving theorems, but it also advances by developing new methods to prove them. Finding these new techniques is part of what mathematics is all about.

They open up new territory for investigation and help us see difficult technical results as manifestations of some deeper truth.

What some mathematicians do, rather than simply prove theorems, is to develop such techniques. The most famous is calculus, developed by Newton and Leibniz in the seventeenth century. Using calculus and his new theory of gravitation, Newton was able to explain in a precise way the motion of the planets round the Sun; not just the shape of the orbits, but the speed of motion. Since then, calculus has been used on a whole range of problems, and is a vital part of mathematics. The character theory that Burnside used is far more specialized than calculus, but it is a very useful technique, and will come up again later as we approach the Monster.

Now let us turn to the table of symmetry atoms that emerged from Dickson's work, and find out how the missing pieces were filled in.

8

After the War

Structures are the weapons of the mathematician.

N. Bourbaki (1935–)

At the start of the twentieth century, in 1901, mathematicians had a table of finite symmetry atoms, given in Dickson's first book, *Linear Groups, with an Exposition of the Galois Field Theory*. But if Lie's groups of transformations provided the prototypes, which they did, then part of the table was missing. Dickson had telescoped things down from the continuous world to the finite world in the classical families *A, B, C*, and *D*, but there were also the exceptional families. Dickson later had some partial success with these, but things were not complete.

Dickson's work at the University of Chicago had started in 1900 and continued for thirty-nine years, but his interests gradually took him in other directions. He published eighteen books and hundreds of research articles, and supervised no fewer than fifty-five PhD students, but it was up to someone else to pursue the other exceptional Lie families. In 1939, the year Dickson retired, war broke out in Europe, and pure mathematics had to take a back seat. Governments had more urgent needs for mathematicians, and by the end of the Second World War there was still no solution.

What was needed was a general method that would telescope Lie's groups down to finite versions in all cases, but part of the problem was curvature. The geometric structure of a Lie group is usually curved, and the way to handle curvature is to make flat approximations, just as we do when we draw a map of the world. Killing's methods for classifying the different families of Lie's groups used exactly this method. The flat approximation comes equipped with extra data, similar to lines of latitude and longitude, but when finite arithmetic enters the picture, the map disintegrates into lots of finite pieces and it was not clear what to do. Fifty years later there was still no solution.

The method of obtaining flat approximations uses 'calculus', the branch of mathematics first developed by Newton and Leibniz in the seventeenth century. For example, if you take a curve and a point on the curve, then the flat approximation at that point is called a tangent line. Using calculus you can find the equations of these tangents. The method, learned by generations of mathematics students, uses a limiting process: take a straight line that intersects the curve at the point you want, and at a nearby point. Gradually move this nearby point until the two points coincide. The line will then be a tangent. But the method fails when we move from the real numbers to finite arithmetic, because there is no such thing as a gradual modification. A different approach is needed.

Finding new approaches to such classical mathematics calls into question the method we used to create it in the first place. Should new methods simply develop in an ad hoc manner, or do we need to form a coherent system of axioms and then develop the modern branches of mathematics from there? The latter course of action is similar to the method Euclid used in 300 BCE when he wrote his *Elements* of mathematics; he founded an axiomatic approach to geometry that reigned supreme up to modern times. Is this the way to do modern mathematics?

To some extent the answer is yes, and the man who rode forward to take up the battle for the axiomatic approach was named after a French general from the Franco-Prussian war (the war that broke out in 1870 and forced Lie and Klein to abandon their stay in Paris). The general's name was Bourbaki, and in the mid-twentieth century his namesake worked assiduously with his collaborators to create a series of books called *The Elements of Mathematics*.

These books developed an abstract and logical approach to various branches of the subject. The attitude behind them is illustrated by an early Bourbaki paper, published in March 1949, and addressed to mathematical logicians.

> I am very grateful to the Association for Symbolic Logic for inviting me to give this address – an honor which I am conscious of having done very little to deserve. My efforts during the last fifteen years (seconded by those of a number of younger collaborators, whose devoted help has meant more to me than I can adequately express) have been directed wholly towards a unified exposition of all the basic branches of mathematics, resting on as solid foundations as I could hope to provide.*

This is written in beautiful English, and although Bourbaki's younger collaborators were all French, and their books were written in French, many of them had worked in the USA, and some still did. In this paper, Bourbaki gave his affiliation as the University of Nancago, a combination of the University of Nancy and the University of Chicago, where its authors were working.

As you may have guessed, Bourbaki was a pseudonym being used by a small cabal of French mathematicians who wanted to create a new axiomatic approach to the main branches of mathematics. Their work started in the early 1930s, and Armand Borel, a later contributor to the Bourbaki project, describes the early years in the following terms:

> In the early thirties the situation of mathematics in France at the university and research levels . . . was highly unsatisfactory. World War I had essentially wiped out one generation. . . . Little information was available about current developments abroad, in particular about the flourishing German school (Göttingen, Hamburg, Berlin), as some young French mathematicians were discovering during visits to those centres.*

The First World War affected France particularly badly because so many of its young mathematicians were sent to the front lines and died there. A wartime directory of the most prestigious establishment for mathematics in France shows that about two-thirds of the student population was killed in the war. In Germany, by contrast, young mathematicians . . . were often employed on scientific work, and their survival reinvigorated the universities when peace returned. Henri Cartan, one of the Bourbaki founders and son of the famous Élie Cartan, whom we met in Chapter 5, wrote:

> After the First World War there were not so many scientists, I mean good scientists, in France, because most of them had been killed. We were the first generation after the war. Before us there was a vacuum, and it was necessary to make everything new. Some of my friends went abroad, notably to Germany, and observed what was being done there. This was the beginning of a mathematical renewal.*

The original Bourbaki group started with two young mathematicians. One was Henri Cartan and the other was André Weil, who became one of the greatest of twentieth-century mathematicians. In 1934 they were both assistant professors at the University of Strasbourg and they found the textbook for one of the main courses inadequate in many ways. Cartan was constantly asking Weil about the best way to present a given topic to the class, so much so that Weil eventually nicknamed him 'the grand inquisitor', and that winter he suggested they write themselves a new text. As Borel writes, 'This suggestion was spread around, and soon a group of ten mathematicians began to meet regularly to plan this treatise.' The loose circle of original contributors met at the Capoulade, a café in the Latin quarter of Paris, to plan the book. The original project was rather naive. They believed they could draft the essentials of mathematics in a few years. In summer 1935 they met for their first congress, but it took four years to produce the first chapter.

The Bourbaki meetings, all held in France, were exceptional. Borel, whom I quoted earlier, recalls his astonishment at the argumentative nature of the process: 'I was not prepared for what I saw and heard, "Two or three monologues shouted at top voice, seemingly independently of one another" is how I briefly summarized for myself my impressions of that first evening.'* Indeed, Jean Dieudonné, who, along with Cartan and Weil, was one of the founding fathers of Bourbaki, confirms Borel's impression:

Certain foreigners, invited as spectators to Bourbaki meetings, always come out with the impression that it is a gathering of madmen. They could not imagine how these people, shouting –

sometimes three or four at the same time – could ever come up with something intelligent.*

Bourbaki meetings were a kind of organized chaos, but it worked, and books were written, one after the other. As to the name Bourbaki, André Weil explained that he and a few other collaborators were in Paris one year when there was a spoof on the annual lecture that first year students were recommended to attend. On this occasion an older student, disguised with a fake beard and an unrecognizable accent, gave the talk. He delivered a clever lecture in which all the theorems were wrong, in various non-trivial ways, although some students claimed to follow the whole thing. His final and most extravagant result was called Bourbaki's theorem. Weil and his collaborators were sufficiently amused to choose the name, which was of Greek extraction, and Weil's wife chose the first name Nicolas.

Had Bourbaki been a practical scientist with a large laboratory to fund, and massive research grants to apply for, he would undoubtedly have done what all such people do. He would have written research papers with his younger collaborators, giving his name as first author. But in mathematics this is not the style. Most mathematics papers are written by a single author. If two people work in collaboration then they publish jointly, but their names are normally listed in alphabetical order. It makes no difference who is senior or junior, who is the original instigator of the research, or who produced the key idea that cracked the problem. If they work together they are considered equal partners. Of course if someone works alone then the paper has only one author, and this remains true even though he or she received advice and help from others. For example, a young person working under a senior mathematician, such as a thesis supervisor, merely writes a note of thanks to the supervisor and anyone else who may have helped with various ideas and suggestions.

The youngest of Bourbaki's original collaborators was named Claude Chevalley, and he certainly published plenty of research papers on his own. In one of his most important, published in 1955, he finally cracked the problem of telescoping all Lie's groups down to finite versions.

Chevalley was born in South Africa when his father was French consul in Johannesburg; he studied in France, and in 1931 went to Germany to continue his studies. He returned to teach in France in 1936, and in 1938 went to the USA. When he wished to return home in the mid-1950s, some mathematicians in France started a campaign against him, since he had spent the difficult years during and after the war in very favourable conditions. However, the campaign was unsuccessful, and he returned to France and worked in Paris until his retirement in 1978. I don't know how Chevalley felt about his detractors, but he was a man who delighted in abstract principles, sometimes to the detriment of practical matters. For example, in 1968 when the Paris students rioted, one contemporary recalls that 'Chevalley took the side of the students and advocated crazy things: every student should not have to take exams (exams are repressive). Yet he had a high sense of quality in mathematics and was extremely demanding for his own students and himself.' His abstract attitudes made him perhaps the most austere of the original Bourbaki group, but Chevalley could produce wonderful mathematics papers and the 1955 paper was a case in point.

Chevalley had finally broken the dam. He had telescoped all

Lie's groups down to finite versions, and other people then swept downstream. One was a young Belgian mathematician named Jacques Tits – we shall hear more of him later – and another was Robert Steinberg in California. They both discovered new families of symmetry atoms inside those that Chevalley had created, and Steinberg gave his paper the title 'Variations on a Theme of Chevalley'.

At the same time, a Japanese mathematician named Michio Suzuki, working at the University of Illinois at Urbana, made a remarkable discovery. He had been working on symmetry atoms having a certain type of cross-section (more on this technique later), and discovered a whole new family of them. His work was independent of Chevalley's, and the Suzuki family at first seemed quite different from those that Chevalley had derived from the Lie families. But they were in fact related, and a Korean mathematician named Rimhak Ree, working at the University of British Columbia in Canada, found the connection. He found three new sub-families among some of the smaller Chevalley families, and one of these turned out to be Suzuki's. Ree's work completed the discovery of all but finitely many symmetry atoms, though no one could be certain of that at the time.

These new families of symmetry atoms were just waiting to be discovered, and when they were, they emerged together as if springtime had just arrived. This is a strange phenomenon in mathematics, and Gauss commented on it: 'Mathematical discoveries, like springtime violets in the woods, have their season, which no human can hasten or retard.' After these new discoveries some experts guessed there were no more families of finite symmetry atoms to be found, at least not with infinitely many members, but it was not clear why this should be the case. If there *were* other families, mathematicians had to find them – and if there were not then they needed a proof of that fact. Such a proof *did* gradually emerge, but only in a long series of very technical papers by many different authors. In the process it turned out that although there were no other families, there *were* some unexpected exceptions. Finding all these exceptions and showing there were no more was the challenge, and it is from this great work that the Monster finally emerged.

The methods that Chevalley, Steinberg, Suzuki, and Ree had used were algebraic. But Jacques Tits in Belgium had used geometry, and some people in other branches of mathematics wondered whether a more geometric approach wouldn't be preferable for all these Lie-type families. After all, Dickson had shown that the classical ones (families *A* to *D*) could all be obtained geometrically, and there ought to be a uniform geometric way of getting the others. There was, and Tits had been working on it for some time.

9

The Man from Uccle

> Symmetry, as wide or narrow as you may define its meaning,
> is one idea by which man through the ages has tried to com-
> prehend and create order, beauty, and perfection.
>
> Hermann Weyl (1885–1955)

Each technical subject has its own specialized vocabulary. Medicine uses terms derived from the classical languages Latin and
Greek. Physics invents its own words, such as protons, neutrons,
quarks, and lasers. Mathematicians need an enormous number of
technical terms; some come from ancient Greek, but others are
quite new, and are often ordinary words with technical meanings
peculiar to the subject.

I was once talking about this to a medical cousin and he
thought it was absurd. Why couldn't mathematicians invent new
terms, like medical specialists do, using Latin or Greek terminology, every time they need them? Surely common words can
be misinterpreted. But mathematicians use *masses* of technical
terms, far more than medical people, and when they need a word,
they give it a definition, rather like Humpty Dumpty in *Through
the Looking Glass* (whose author was an Oxford University
mathematician).

Some words acquire a special status and become standard

terminology, impenetrable to anyone outside the subject. For instance, mathematicians use the word 'building' in a *completely* different way from its meaning as a place for living or working; it refers to a mathematical object built up from crystal-like structures. Each finite symmetry atom in the 'periodic table' has its own building, and buildings provide a geometric explanation for all symmetry atoms in the table, in contrast to the algebraic perspective of Chevalley and others.

Jacques Tits invented mathematical buildings, though he didn't call them that at first. Tits was born on 12 August 1930 in Uccle, an ancient town that is now a suburb of Brussels. As a child of three or four he was a mathematical prodigy, who surprised visitors by being able to do all the operations of arithmetic before practically being able to speak. He started school at the usual age of six, but soon skipped a year, and later skipped several more.

Tits's father was a mathematician: 'He explained many things to me, but after a while he refrained from explaining too much.' This didn't stop the young Jacques: 'I took the books from his bookshelves and started to read them.' But suddenly there was an abrupt change of circumstances. 'My father died when I was thirteen, and my mother then had very little to live on. The teacher at school knew all this, and he suggested tutoring people to contribute to the household expenses. I started tutoring people four years older than me who were going to study engineering at university.' A year later the young Tits himself went to university. The sooner he entered, the sooner he would be able to earn a salary and help to support his mother. He passed the entrance exam at fourteen, started his studies at the Free University of Brussels, and continued tutoring to earn money. Brussels, as the capital of Belgium, now has two free universities: a French one (the Université Libre), and a Flemish one (the Vrije Universiteit), but at the time there was only the French one, which was fine because French was his mother tongue, though his surname was Flemish. In 1950, when still only nineteen, he obtained his PhD.

Tits had a very geometric turn of mind, and in the early 1950s he was working to develop a more geometric approach to Lie's groups of transformations. As we saw earlier, Lie's groups came in seven families, A to G, and Dickson had used the geometry under-lying families A, B, C, and D, along with $E6$ and $G2$, to obtain finite versions. Tits wanted to do the same thing for all families, and create finite versions of Lie's groups in all cases.

Unfortunately for Tits, the more experienced mathematician Claude Chevalley, who was over twenty years older, was already on a similar track, using algebra rather than geometry. 'I was working with geometric ideas, but Chevalley had a faster method in which he had a general construction, which I did not have at first.' Tits and others soon produced variations on Chevalley's theme, but Tits was also working on a new theory that would include all these variations. It took years to reach a fully developed form – a new mathematical theory needs time to mature, like a good wine – but when the theory of buildings was ready it was welcomed by discriminating judges of good mathematics, such as the Bourbaki group. More on that later, but what on earth is a building?

Buildings are multi-crystals in a sense I shall explain, and I would love to draw a picture of one, showing all its glorious symmetry. Unfortunately this isn't possible. A multi-crystal lives in a collec-tion of intersecting universes, and a picture in our own universe is always squashed out of shape. If one part is right, another is wrong, and most of the symmetry is lost. Pity. But let us look at the simplest case, based on a flat two-dimensional crystal. Take a hexagon. It has six edges, the word *hexa* coming from the ancient Greek word for six.

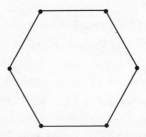

A building, or multi-crystal, built out of hexagons is a network of edges satisfying two conditions: there can be no circuit of fewer than six edges; and any two edges must lie on at least one common hexagonal circuit.

Here is a simple example. In this picture there are three paths of length three from the top vertex to the bottom one; any two of these paths, taken together, form a hexagonal circuit. This gives three circuits, and each pair of edges lies on at least one of them.

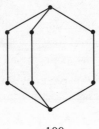

This example is too simple to show much symmetry. If we go for lots of symmetry then each vertex must lie on the same number of edges, as in the next picture, where each vertex lies on exactly three edges. This is a pleasing picture, but every single hexagon is pushed out of shape and most of the symmetry is lost.

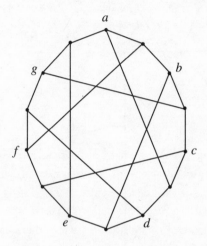

Here is how to read this picture. Treat it as a network of fourteen vertices, all on the outer circuit, joined by twenty-one edges, fourteen on the outer circuit and seven longer ones crossing it (ignore the points where the interior edges cross one another). Although the drawing is quite elegant, it fails to exhibit much of the symmetry: in a perfect geometric realization all edges would have the same length and all hexagonal circuits would be perfect hexagons, but it is not possible to draw it that way. There are twenty-eight hexagonal circuits, and in this picture they come in four different shapes, shown below. There are seven of each.

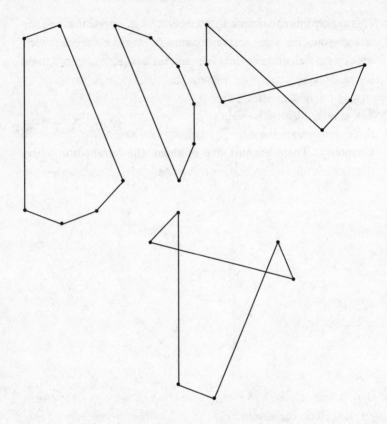

The symmetry of this multi-crystal is not self-evident. It arises from permuting the fourteen vertices among one another in a way that sends edges to edges. In other words, if a pair of vertices is joined by an edge, then after each permutation they must still be joined by an edge. The group of these permutations is what I am referring to when I talk about the symmetries of this multi-crystal.

If you think this looks complicated, then you are in good

company. Mathematicians find multi-crystals impossible to visualize, so they just view parts of them. For example, they may view a single crystal, and use their imagination and some algebra to do the rest.

The crystals themselves follow a well-understood pattern, and in three dimensions they are the regular polyhedra that we saw in Chapter 1. There are just five of them: the tetrahedron, cube, octahedron, dodecahedron, and icosahedron, shown below.

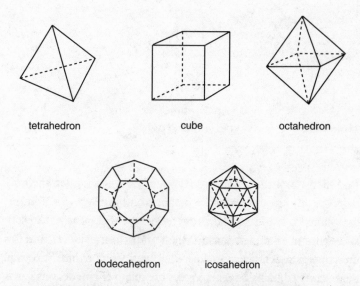

tetrahedron   cube   octahedron

dodecahedron   icosahedron

A *multi*-crystal based on one of these polyhedra is far too complicated to be shown in a picture. It has lots of faces, and for each pair of faces there is a crystal containing both. No one tries to picture the whole thing, but this doesn't stop mathematicians from working with them – and not only in dimension 3. In order to get all finite symmetry atoms in the table, Tits needed crystals in dimension higher than 3, and then had to combine these higher dimensional crystals into multi-crystals.

Higher dimensional *multi*-crystals sound pretty horrendous, but the *single* crystals in higher dimensional space are not nearly as diabolical, as you may judge for yourself.

Dimension	Types of crystals				
3	$A3$	$B3$			$H3$
4	$A4$	$B4$		$F4$	$H4$
5	$A5$	$B5$			
6	$A6$	$B6$	$E6$		
7	$A7$	$B7$	$E7$		
8	$A8$	$B8$	$E8$		
More than 8	Types A and B only				

In dimension 3 there are three types of crystals: the tetrahedron has type $A3$, the cube and the octahedron have type $B3$, and the dodecahedron and icosahedron have type $H3$. The table lists analogues in all dimensions for the tetrahedron (type A), and for the cube and octahedron (type B). I want to convince you that these higher dimensional analogues are relatively innocuous – not trivial, of course, but not diabolical either.

Type A in two dimensions is the triangle, which has three vertices. In three dimensions it is the tetrahedron: this has four vertices and every pair of vertices forms an edge. In four dimensions there are five vertices and each pair forms an edge. In five dimensions there are six vertices, and so on. In all cases each pair of vertices forms an edge. Of course there are triangular faces, (solid) tetrahedral faces, and so on, but the underlying structure is quite simple.

Type B in two dimensions is the square. In three dimensions it is the cube (or the octahedron, but let's concentrate on the cube). To view its four-dimensional analogue, first view the cube in perspective, face on. In the picture the big square is the front face and the small square is the back face. The other four faces lying in between appear deformed out of shape by the perspective.

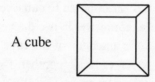

A cube

In the four-dimensional analogue of the cube, the square faces go up a dimension to become cubes. Instead of seeing one square inside another we see one cube inside another. Here is the picture.

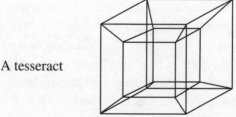

A tesseract

Once again we are looking at things in perspective. The four-dimensional cube (which is called a tesseract – I'll explain why below) has been projected into three-dimensional space, and then of course into two dimensions to get it on to a sheet of paper. The big cube and the small cube are the front face and the back face of this tesseract. The in-between pieces are the other (cubical) faces. There are six of them. In four dimensions they would be perfect cubes too, but the perspective renders them out of shape. There are computer simulations of four dimensions showing the tesseract rotating. As different faces move to the front, they become perfect cubes.

This tesseract is a four-dimensional crystal of type B, and its name comes from the Greek word *tessera*, meaning four. If you were measuring four-dimensional volumes you would use 'tesseractic centimetres', as opposed to cubic centimetres in three dimensions, or square centimetres in two dimensions.

In five dimensions the analogue of the cube and tesseract can be pictured as one tesseract inside another. Corresponding corners have to be connected by edges. Call it a pentact, then go to six dimensions by having one pentact inside another, and so on. Crystals of type B exist in all dimensions.

As the table shows, there are other types of crystals in dimensions 3, 4, 6, 7, and 8; and the first person to discover the ones in dimensions 6, 7 and 8 was a lawyer, not a mathematician! They were related not to a legal case, but to the *absence* of a legal case. A young lawyer named Thorold Gosset in late nineteenth-century London had time on his hands during his early years, and amused himself with investigating higher dimensional space. In 1900 he published his results. We shall come back to Gosset's exceptional crystals later.

Multi-crystals, as I mentioned, were the invention of Jacques Tits. His work attracted a lot of attention and he moved from Brussels to the University of Bonn. In 1974, the same year his book on this subject finally came out, he then took up a position at the Collège de France, a prestigious research establishment in Paris. In the book, Tits proved a remarkable theorem that when the crystals themselves are at least three-dimensional, then the whole multi-crystal has immense symmetry. He used this theorem to help find all these multi-crystals, giving a very pleasing geometric explanation for the symmetry atoms in the 'periodic table'.

This was elegant mathematics, but since these multi-crystals are so complex, you may wonder how to construct them. Tits showed how to get them from the symmetry atoms, which in turn can be constructed using the algebraic methods of Chevalley. But this is a slightly roundabout method, and it is preferable to construct the multi-crystals first and then obtain the symmetry atoms later, giving a more geometric approach.

As so often in mathematics the solution was found by first dealing with a slightly different question. In 1984, Tits gave a lecture at a conference in the mathematical research institute at Oberwolfach, a wonderful establishment located in the Black Forest region of Germany. It was once a hunting lodge, but was turned into a mathematics institute in 1944. After the Second World War, its director, Wilhelm Süss, used his connections to obtain funds for its rebuilding and expansion. It is now arguably the finest place in the world to conduct a mathematics conference. Tits's lecture was on multi-crystals in which each 'crystal' was a plane, tiled – like a bathroom wall – by equilateral triangles (the plane extends to infinity in all directions, so these multi-crystals are infinite and not directly related to finite groups of symmetries). I had recently been thinking of such multi-crystals and found a way of growing them, starting at one vertex and moving outwards. The trick was to attach a small 'genetic code' to each vertex showing how it must grow around that vertex, like the petals of a flower. This does not mean using a laboratory with expensive equipment and a carefully controlled environment, but just refers to a theoretical construction.

The following summer, Tits and I worked in Paris to adapt this method to multi-crystals in which the crystals were polyhedra, like the octahedron, rather than infinite tilings of the plane. This was a harder problem because the faces of a polyhedron have to join up as you go around the back of the polyhedron, unlike a tiling of the plane that goes off to infinity. But Tits had an idea how to tackle it, which he explained to me and another mathematician named Pierre Deligne, who had taken university courses from Tits while still a teenager at high school in Brussels.

This is how mathematics is done. People sit and talk, perhaps with a chalkboard at hand, and as they talk they clarify their own ideas. Deligne was a good person to talk to because he was becoming one of the century's greatest mathematicians, and Tits once remarked, 'He is amazing. You explain something to him and within two minutes he understands everything you know about it and much more besides.' We were sitting in a ground floor office on a sunny day, and as Tits tentatively wrote a few symbols on the board, Deligne interrupted with an objection. He immediately saw beyond the solution to an intriguing complication that was important later – I was amazed.

In the end we found that by combining the 'genetic codes'* for growing the three-dimensional multi-crystals, we could grow multi-crystals in all dimensions greater than 3. Geometry, rather than Chevalley's algebraic methods, could now yield almost all symmetry atoms. It is remarkable that three dimensions were enough, but it seems to happen frequently in mathematics that once you figure things out at a lower level, the higher levels take care of themselves. Our 'genetic codes' were elementary, yet we were able to create complicated objects such as multi-crystals of type $E8$, where even the simplest one has more faces than there are particles in the universe.

Before we leave multi-crystals, it is worth noting that they give rise to all sorts of fascinating patterns. Here is one that will give the idea. In the multi-crystal pictured on page 101, seven alternate vertices on the outside circuit are labelled with the seven letters a to g. Each of the other seven vertices on the outside circuit is joined to three alphabetic ones. This gives the following seven blocks, each having three symbols.

abf
bcg
acd
bde
cef
dfg
age

These blocks can be used to label the other seven vertices of the multi-crystal, but let us forget about multi-crystals for a while. If you look carefully, you will notice that each pair of symbols lies in exactly one block. And each pair of blocks has exactly one symbol in common.

This is a remarkable pattern, and one can ask whether it is possible to do a similar thing with four symbols per block. Each pair of symbols is to lie in exactly one block, and any two blocks are to have a symbol in common.* The answer is yes. With four symbols per block you need 13 symbols and 13 blocks.

Patterns like this have useful applications. For example, suppose you wanted to conduct a series of experiments in which each pair of subjects is included in exactly one of the experiments, and any two experiments share one subject. Represent each subject by a symbol, and each experiment as a block of symbols.

Can you do this with blocks of larger size? The answer is yes for block size 5 and block size 6, but not for block size 7. It is impossible for block size 7; not difficult, but impossible. This can be proved – no ifs, ands or buts. But for block sizes 8, 9, and 10 it *is* possible, so what's going on?

Some block sizes are possible and some are impossible. For the sake of argument let us call the block size $q + 1$ – you will see why in a minute. We want a collection of blocks, each of this size, having the property that every pair of symbols lies in a block, and every pair of blocks has exactly one symbol in common.

This is possible if q is a prime number, or a power of a prime number. In other words it is possible if q is 2, 3, 5, 7, 11, etc. (the prime numbers), or if q is $4 = 2^2$, $8 = 2^3$, $9 = 3^2$, etc. (powers of prime numbers). On the other hand, it is not possible when q is 6, meaning block size 7 is impossible. After 6, the next problem number is 10. When q is 10 the blocks would have size 11 and there would be 111 symbols.

Does such a block system exist? In the late 1950s this was already an old problem, and some began to think that computers might be used to resolve it. This finally happened in the 1990s, by which time computers were vastly more powerful, but computer proofs are unsatisfactory because you can't check them by hand. We will have more to say about this later.

After 6 and 10, the next problem number is 12. Someone who is a master at creating strange objects, and will appear later in our story, tried to construct an example when q is 12. He tried very hard, for years, using good ideas and powerful computers. But even he gave up. It would not surprise anyone to learn that q has to be a prime number or a power of a prime number. If I were a betting man, I would bet on it. But I wouldn't bet much because it may be false. And that is the trouble with mathematics. No amount of evidence will do it. You have to *prove* things, one way or the other. It can be a frustrating process, to say nothing of the technicalities and abstractions that put so many people off the subject. The necessity to provide proofs prompted one fellow mathematician, who moved far into the stratosphere of university administration, to comment to me once, 'It's a brutal subject.'

At the start of this chapter I referred to multi-crystals as buildings, but where does the term 'building' come from? Jacques Tits, who invented them, used other terminology to start with. He had been led to create buildings by first studying geometry in the more traditional sense involving points, lines, and planes, and he continued using the term geometries (with a suitable definition of what exactly he meant) for his new creations. The term 'building' was first used by Bourbaki. Remember him, the reincarnated French general who took up the task of laying out the basics of higher mathematics? Of course, Bourbaki was writing in French so he didn't use the word building, but the French word *immeuble*; building is the English translation. But why did Bourbaki choose this word?

Tits's multi-crystals were an amalgamation of crystals. These crystals arose in a natural way in the Lie theory, where their faces were called 'chambers'. Tits once called the crystals 'skeletons' because they are the bare bones of the subject, but this mixes metaphors. With the word chamber in mind, Bourbaki renamed the crystals 'apartments', and called the whole thing a building.

10

The Big Theorem

Reductio ad absurdum [a method of proving theorems],
which Euclid loved so much, is one of a mathematician's
finest weapons. It is a far finer gambit than any chess play: a
chess player may offer the sacrifice of a pawn or even a piece,
but a mathematician offers the game.

G. H. Hardy, *A Mathematician's Apology*

In his novel *Uncle Petros and Goldbach's Conjecture*, Apostolos
Doxiadis portrays a fictional mathematician who spends his
whole career trying to settle an old question. Is every even
number a sum of two prime numbers? This happens to be
true for all even numbers that anyone has ever looked at, and
with the help of computers they have checked everything up to
60,000,000,000,000,000. But how do you prove it for *all* even
numbers? No one knows.

Most pure mathematicians, like the fictional character in the
book, would love to prove a really hard theorem. But what makes
a theorem hard? Is it a conjecture like this that no one knows
how to solve, or is it something whose proof is necessarily very
complex, like a Mount Everest that no one can climb without
setting base camps high up on the slopes and taking suitable
equipment and warm clothing? The Goldbach conjecture – every

even number is a sum of two prime numbers – certainly lies in the first category, and probably in the second too. But another statement that refers to even numbers, and certainly lies in the second category, is a theorem in the mathematics of symmetry (that branch of mathematics called group theory). It makes a deceptively simple claim: apart from the prime cyclic groups, the size of every symmetry atom must be even. A symmetry atom, remember, is a group of symmetries that cannot be deconstructed into anything simpler, and there is another way of stating this theorem: if a group of symmetries has odd size, then it *can* be deconstructed (and a complete deconstruction will yield a collection of prime cyclic groups). Stated in this form it is sometimes called the 'odd order theorem'.

This theorem is vitally important, as I will explain later, and it set in motion a long sequence of results that led to the discovery and classification of *all* finite symmetry atoms. Its proof, by Walter Feit and John Thompson, places it firmly in the Everest category.

Now you would think that a result of immense importance like this, and one that could be stated so simply and directly, would be welcomed with open arms by any and every mathematics journal in the world. Not so. Several journals declined it on account of its length. Weighing in at 255 pages of carefully reasoned argument, it was too big for them. Ten pages, 20 pages, even 40 or 50 pages for a really major result, was normal and acceptable, but 255 pages? The paper took up one whole issue of the *Pacific Journal of Mathematics*, whose editors were immensely proud to publish it.

When a mathematician submits a paper to a journal, and the editors consider it for publication, they first send it to a referee for comments. Some referees, before recommending acceptance, read every last word of the paper, submitting detailed notes to the

114

editors, correcting infelicitations in the prose and opacity in the technical details, and sometimes suggesting shortcuts. In the case of the Feit–Thompson paper, it would be quite unreasonable to expect this, but then it is not an obligation of the referee. It would be embarrassing if the result turned out to be wrong, but the real embarrassment would be for the author, or authors. Who were these authors, Feit and Thompson, and why was the theorem so important? Let us deal with the second question first. This takes us to the work of a German mathematician who moved to North America in 1933.

On 30 January 1933, Hitler and the Nazi party took over the government of Germany, and on 7 April 1933 the new Civil Service Law expelled most Jewish academics from their jobs (to be classed as Jewish, or more precisely non-Aryan, it was sufficient to have one Jewish grandparent). Richard Brauer, a young man of 32, had been a mathematician at the University of Königsberg (now called Kaliningrad) for eight years already, but with the Nazi party's new edict he lost his job and had to find employment abroad quickly. He went to the USA, while his elder brother Alfred, who had first inspired Richard with a love of mathematics, retained his job at the University of Berlin and remained in Germany. There was an exemption clause for those who had fought for Germany during the First World War, but decisions at the Nuremberg party congress in autumn 1935 enabled the authorities to ignore the law, and Alfred too lost his job.

What happened in 1933 was quite phenomenal. A young American mathematician named Saunders Mac Lane, who will reappear in due course, decided to go to Germany to study. He went to the mathematical powerhouse of Göttingen, after 'spending a vaguely disappointing year of graduate study at Chicago'. He arrived in 1931, and found a very stimulating atmosphere, but soon witnessed the devastation caused by the anti-Jewish measures in 1933. On 3 May, Mac Lane wrote to his mother: 'So many professors and instructors have been fired or have left that the mathematics department is pretty thoroughly emasculated.'* When Issai Schur, Brauer's thesis supervisor at the University of Berlin, was fired in 1933, there was horror among his colleagues, and as a contemporary wrote, 'When Schur's lectures were cancelled there was an outcry among the students and professors, for Schur was respected and very well liked.' But as time went by people accustomed themselves to the Nazi edicts. Schur could not understand the persecution and humiliation that was heaped on him, and as one contemporary reported later, referring to his sixtieth birthday in January 1935:

> Schur told me that the only person at the mathematical Institute in Berlin who was kind to him was Grunsky, then a young lecturer. Long after the war, I talked to Grunsky about that remark and he literally started to cry. 'You know what I did? I sent him a postcard to congratulate him on his sixtieth birthday. I admired him so much and was very respectful in that card. How lonely he must have been to remember such a small thing.'*

Yet prior to 1933 the German universities were great beacons of light. This was where the Bourbaki mathematicians went to learn about new developments after the First World War. The German university system was the most stimulating place in the world at that time, and as Richard Brauer himself wrote later in life, 'The intellectual atmosphere of German universities of that period is remembered with nostalgia by all who knew it.'* There was always the hope that the Nazi menace would be a brief aberration and things would settle down. They didn't, and Richard's sister, who remained in Germany, was killed in an extermination camp during the war.

In the meantime, Richard Brauer found a one-year position at the University of Kentucky in 1933, and then went to the Institute for Advanced Study in Princeton for a year. The Institute had just been founded in 1930, and Brauer was assistant to the great Hermann Weyl, also an expatriate German, and was thrilled to be working with him: 'I had hoped from the day of my PhD thesis to get in touch with him some day; this dream was now fulfilled.'* Weyl was very keen on the relation between mathematics and physics, and one of his collaborative papers with Richard Brauer provided a mathematical background for the concept of electron spin in quantum mechanics.

Next year Brauer took up a position at the University of Toronto, where he remained for the next 13 years. In 1948 he moved to the University of Michigan, and in 1952 to Harvard. Richard Brauer gives the lie to the notion that mathematics is always a young person's game. He was 51 when he was appointed at Harvard, and as one of his biographers writes, 'It is a striking fact of his career that he continued to produce original and deep research at a practically constant rate until the end of his life.'* Nearly half of his publications were written after he was 50, and they include one that gave a means of finding all finite symmetry atoms, and stimulated the Feit–Thompson theorem.

The point is this. If a symmetry atom has even size, then a theorem of Cauchy – he who mislaid some of Galois's papers – guarantees that it contains a symmetry of order 2. This means a symmetry that when done twice leaves everything as it was. For example, a mirror symmetry has order 2: do it once and everything is reversed across the mirror; do it twice and everything is back to where it was. Within a symmetry atom, a mirror symmetry is a very small thing, but it has big consequences. Most of the symmetries will move the mirror to another mirror, but some will stabilize it. The sub-group of symmetries stabilizing the mirror I shall call a cross-section, though mathematicians have a more technical term for it.*

A symmetry atom cannot have a single cross-section, because it can operate as a group of symmetries on itself, moving a cross-section into many different positions. This yields lots of cross-sections having the same shape, and from these cross-sections you can reconstruct the symmetry atom, just as you can reconstruct a picture of the brain using cross-sectional brain scans. Now the vital thing that Brauer proved, along with his student K. A. Fowler, was that there are only a limited number of ways that cross-sections of the same shape can fit together. This means there are only a limited number of symmetry atoms with the same cross-section, and very often there is only one. In other words, once you know a cross-section you have almost nailed the symmetry atom. Brauer went on to prove that for the symmetry atoms in some Lie families, a cross-section uniquely determined the whole thing. This was terrific stuff because it meant there was a possible way of finding all symmetry atoms.

Here is the idea. Take the symmetry atoms you already know about and look at their cross-sections. In each case show that there are no other symmetry atoms having such cross-sections: either you prove that, or you find new symmetry atoms. Do the same for other possible cross-sections: either show they cannot occur in any symmetry atom, or find new symmetry atoms. If you find any new ones, then you consider them as possible cross-sections in something larger. This is *precisely* how the Monster was discovered, by first discovering one of its cross-sections, but more on that later in the book. When this process eventually draws to a close, you have a list of *all* symmetry atoms.

This was a wonderful plan. There was just one glitch. How did we know that all symmetry atoms had cross-sections? Or what amounts to the same thing, how did we know that the size of a symmetry atom had to be an even number? This is precisely what Feit and Thompson proved. It was a huge result, and Daniel Gorenstein, who later orchestrated the programme of discovery and classification of all symmetry atoms – the great 'Classification' project – writes, 'The single result that, more than any other, opened up the field and foreshadowed the vastness of the full classification [of all symmetry atoms] was the celebrated theorem of Walter Feit and John Thompson.'* But where did Feit and Thompson come from and how did they get to their theorem?

When the Nazi government forced Richard Brauer out of his job in 1933, Walter Feit was just three years old, and living in Vienna. By 1939 his parents decided to place him on a KinderTransport that carried Jewish children to safety. He left Vienna on 1 September 1939. His parents had plans to follow a fortnight after, but two days later, on 3 September 1939, the Second World War started. He never saw them again. As young Walter arrived in England, to stay with an aunt in London, the British government was already evacuating children to more rural areas from the capital city. He was relocated several times, but eventually won a scholarship to a school in Oxford, where he stayed until after the war.

In late 1946, Walter left school and went to the USA, where two days after his arrival he was whisked into a family gathering of over 400 people in New York. The next day he wrote back to his aunt in London, reflecting optimism and happiness in his new surroundings.

Dear Auntie Frieda,

I have hardly had a moment to spare until now . . . Yesterday night was New Year's Eve in the USA. This is an important holiday so I went to the [family] banquet. There were over four hundred people present. . . . On Monday I was outfitted for Miami; I now possess five new pairs of trousers, two new jackets plus new shoes . . . I also have a watch in my possession. . . . There are many things in this country that I shall have to get used to. For instance, I am now sitting in the kitchen of my uncle. He is not especially well off, yet I can see a huge refrigerator, artificial daylight, an electric clock . . . and several other things including central heating such as are never seen in an ordinary English flat.*

Miami was his next stop, where he stayed with an aunt and uncle, and the following September he went to study at the University of Chicago.

Like many who escaped from the Nazis, Walter Feit suppressed stories of his early life, and left his past behind. But in 1990, at a conference in Oxford honouring his sixtieth birthday, he astonished his audience by telling them he had been educated at a school in Oxford, before becoming a student in Chicago.

At the University of Chicago, Feit was awarded a master's degree, and received his bachelor's a week later (not the usual order of events!) From there he went to the University of Michigan to work under Richard Brauer. When Brauer died, a quarter of a century later, Feit wrote an elegant and very informative obituary of him for the American Mathematical Society. It deals with the German years, and the events that led to his leaving for America, but from the way Feit writes one would never know that he himself made a narrow escape from the terror that followed. The KinderTransport that took him to safety was the last one to leave Vienna.

When Brauer moved to Harvard, Feit stayed in Michigan to finish his PhD, then took a job at Cornell. Being only 22 at the time, he was soon called up for service in the US Army, but on his return to Cornell 18 months later he was contacted by the young John Thompson.

Thompson had been an undergraduate at Yale, where he started by studying theology. After the first year he switched to mathematics and did extremely well, and Mac Lane invited him to come to the University of Chicago for graduate study. Thompson became interested in finite mathematics (finite symmetry groups), which was not a fashionable subject in those days, and one eminent faculty member expressed doubts about him: 'Be careful of this John Thompson fellow – I'm not sure he's reliable.' Attitudes of the professors rub off on the graduate students and some of them pinned up a witty poem making fun of finite mathematics. However, Saunders Mac Lane, who did *very* fashionable mathematics, and sported tartan ties and jackets, took on the young John Thompson as a PhD student. Mac Lane realized that finite group theory was an immensely technical subject, and said he didn't feel capable of doing it himself, so it was rather brave of him to take on a student in this area. But Thompson was a self-starter and an independent thinker, and when Mac Lane went away for part of the year in 1958, he confidently left him under the eye of another mathematician who had just arrived.

This new man, Dan Hughes, recalls Thompson doing mathematics at a dizzying rate, 'I still remember these yellow sheets of paper he kept producing. Every day he'd bring in ten or twelve of these.' Thompson was a determined student, and finally cracked an old conjecture that had been around for 60 years. It formed a spectacular PhD thesis, but Thompson, like lesser mortals, still had to take an oral exam. The senior man who had initially expressed doubts was on the committee and said, 'It makes me feel very silly to be examining a guy like this.' But examine him they did, and when it was over he left the room. They delayed long enough for propriety and called him back. 'Well, John, we've talked it over and it's been a hard decision, but we've decided, etc. etc.' As a mere graduate student, Thompson may not have realized they were teasing him.

Once Feit and Thompson got into correspondence, and started collaborating, they aimed for the stars. The idea, as I mentioned above, was to show that the size of every symmetry atom is even. This is equivalent to proving that if a group of symmetries has *odd* size, then it can't be a symmetry atom, and this was the approach Feit and Thompson took. Fortunately for them, Michio Suzuki – he of the Suzuki family of symmetry atoms – had recently dealt with a special case of this problem, and his methods gave them a conceptual framework to work in. Thompson recalls that 'By 1959 we were going at it hammer and tongs', and in collaboration with Marshall Hall at the California Institute of Technology, they extended Suzuki's result to a less specialized case.

In the meantime, Thompson acquired his PhD, and a senior mathematician at the University of Chicago, named Adrian Albert, who had excellent connections to the intelligence establishment, recommended he go to the Institute for Defense Analyses (IDA). He spent the year 1959–60 there, away from the usual academic environment, and when his PhD supervisor Mac Lane returned to the department, he was furious. But Thompson continued his mathematical research at the IDA, and meanwhile at the University of Chicago they decided to arrange a special year on finite mathematics.

This brought Thompson back, and he and Feit intensified their work together. They were using the well-known trick of assuming a counter-example to their theorem, and showing that it would lead to a contradiction. In other words, they took a symmetry atom whose size was an odd number, and tried to show it couldn't exist. They made very sophisticated use of a technique called character theory, which Burnside had used earlier in the century, and Feit was an expert on this important topic, but as Gorenstein says, 'Unfortunately an even greater obstacle awaited Feit and Thompson, for one of the final configurations . . . completely eluded the hoped-for contradiction [and] it was a full year before Thompson found a way of eliminating this last configuration.'* Thompson's method was very technical, concise, and clever. Only an extraordinarily dedicated person could produce such arguments, and as Jonathan Alperin, a colleague of Thompson's at the University of Chicago at the time, said recently, 'Thompson worked every single moment for years. The only other person like that I can think of is Bobby Fischer, the chess player.'

Years later, in 1970, Thompson's work, on this and other things, was recognized by the award of a Fields Medal, the greatest accolade in mathematics, and more enviable than a Nobel Prize. Since it was first instituted in 1936, there have been forty-six recipients of this medal, compared to 134 Nobel Prize winners in physics during the same period. Fields Medals are awarded every four years and for each one a senior mathematician speaks about the recipient's achievements. In Thompson's case the senior mathematician was Richard Brauer, and one of the things he discussed was, of course, the Feit–Thompson theorem:

> Here, the authors proved a famous conjecture, to the effect that all finite [symmetry atoms] have even size. I am not sure who was the first to observe this. Fifty years ago it was already referred to as a very old conjecture . . . [but] nobody ever did anything about it, simply because nobody had any idea how to get started.*

After proving a theorem like this, what do you do for an encore? The whole point of the Feit–Thompson theorem was to open the way to classifying all symmetry atoms, so one obvious thing was to set about this task. In the early days, Thompson was overheard saying he would knock this off in short order, but there was no such luck. It turned out to be very complicated indeed because of the exceptions, which eventually led to the Monster, but more on that later.

Remember what the idea was. You take a cross-section in something you know about, and show that there is nothing else having this cross-section. Brauer had already dealt with some families, but there was still plenty of work to do. After the special year at the University of Chicago, Thompson went to Harvard to finish writing the proof of the Feit–Thompson theorem, and work in the presence of Brauer.

In 1962 Thompson came back to a position in Chicago, and started work on cross-sections of type $A1$. One class of these arose in a special family of symmetry atoms, but the rest were not cross-sections in any known symmetry atom. Thompson wanted to prove that this was the end of the story, so he took an imaginary symmetry atom having a cross-section of type $A1$, and tried to show that either it was in the family he wanted, or it led to a contradiction.

Now you might suppose that cross-sections of type $A1$, rather than those of higher rank like $A2$ or $A3$, ought to be relatively easy. But this is not the case at all. Think of a monocycle, a bicycle, and a tricycle as analogues to type $A1$, $A2$, and $A3$. The monocycle is the trickiest to deal with – and you can do things with it that are impossible with bicycles and tricycles. It is that way in mathematics – the low rank cases are trickiest and that is precisely where unusual things can happen. Thompson worked very hard on the problem, and eventually wrote up his results.

He hadn't yet prepared them for publication, and was working on other things, when in 1964 he received a letter from a mathematician named Zvonimir Janko in Australia. Janko had approached the same problem in connection with his own work, and had found that when the cross-section was the smallest symmetry atom in the $A1$ family he couldn't get the required contradiction. Alperin remembers the occasion: 'I remember it very clearly. Thompson told me of the letter at tea, and he was smiling about it. The next morning he wasn't smiling.'

Thompson had already replied to Janko, but immediately after posting the letter he noticed an error in his own argument. He wrote a second letter and they got into communication, and decided to publish a joint paper, leaving out the awkward case. Janko was already working on it and would continue with his efforts.

11

Pandora's Box

Look around when you have got your first mushroom or
made your first discovery: they grow in clusters.

George Pólya (1887–1985)

In mathematics, as in other creative activities, you can get
completely stuck, unable to move towards a proof of your conjecture, and unable to disprove it by finding a counter-example. In
Chapter 10 we left Thompson and Janko preparing to publish
their results on cross-sections of type $A1$, while Janko continued
working on the one awkward case, where Thompson's original
contradiction had not worked.

He took an imaginary symmetry atom having the awkward
cross-section, and threw all his efforts into obtaining a contradiction to show it couldn't exist. This involved figuring out numbers
associated with the imaginary object, and showing they led to a
contradiction. If there was no mistake in the calculation, then
there was no symmetry atom having the awkward cross-section.
After he got a contradiction, Janko would write down his calculations in great detail and check them. But every time this happened
he found an error in his work and the contradiction melted away.
Could this be because there really was something out there? Janko
thought that might be the case, but every time he tried to work out

more details about this imaginary symmetry atom he got another contradiction.

Was it reasonable that there might be a symmetry atom not in the table? The short answer was yes, absolutely, because five exceptions were discovered in the mid-nineteenth century. But these five all had a very remarkable property, and no one seriously supposed there was anything else quite like them. If there was it would have been discovered already. This suggested that any further exceptions, if they existed, were probably of quite a different nature. Let us leave Janko worrying away at the problem, and go back to the nineteenth century to find out about the five strange exceptions that were already known.

In the mid-nineteenth century, after Galois's work was published, mathematicians gradually got intrigued by groups of permutations, and developed the concept of 'transitivity': a group of permutations on a set of objects was called transitive if it could send any object to any other. This is frequently the case. For example, the group of symmetries of a square permutes its four corners transitively; any corner can be sent to any other. Transitivity is a common occurrence, but there are higher levels of transitivity that are far less common.

If a group of permutations can send any *pair* of objects to any other pair, then it is called 2-fold transitive. This is not so common. With the square, for instance, a pair of vertices joined by an edge is quite different from a pair not joined by an edge. You can't send one pair to the other, so although the symmetry group of the square is transitive on its set of four vertices, it is not 2-fold transitive. For example, in the picture below, a 90° turn will send the pair a,b to the pair b,c but there is no symmetry of the square sending the pair a,b to the pair b,d.

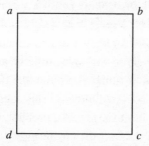

Of course, if you take the group of all permutations, or all even permutations, on a set of objects, then 2-fold transitivity is no problem, but those groups are too large and inclusive. They mix everything around too much and don't preserve an interesting pattern. Are there smaller groups that are 2-fold transitive? And while we are on the topic, what about 3-fold transitivity, 4-fold transitivity, and so on? For example, 3-fold transitivity means being able to send any triple to any other triple.

An example of a pattern whose group of symmetries is 2-fold, but not 3-fold, transitive appeared in Chapter 9. There were seven symbols and seven blocks: any pair of symbols determines a unique block, and the group of symmetries for the whole pattern can send any pair to any other pair. Here is the pattern again:

> *abf*
> *bcg*
> *acd*
> *bde*
> *cef*
> *dfg*
> *age*

The group of permutations preserving this pattern is 2-fold transitive, but not 3-fold transitive. For example, the triple a,b,f forms a block and can only be sent to another triple that also forms a block. You cannot, for instance, send a,b,f to b,c,d – if a goes to b, and b to c, then f must go to g.

Multiple transitivity is rare, and the higher the level of transitivity, the more likely it is that the group will include all permutations, or at least all even permutations. By the time you reach 6-fold transitivity nothing else is possible. This fact has been proved using the list of *all* symmetry atoms,* but as a method of proof this is not very elegant, *and* it only works after verifying that the list is complete. Getting everything on the list and *proving* that it is complete was an immense collaborative effort, like the teamwork that led to the first nuclear bomb. So here we have a mathematical analogue of using a nuclear bomb to crack a very hard nut. It works but one wishes there was another way. If there is, then no one has found it yet.

If we ease up a bit and ask only for 5-fold transitivity, then two very strange beasts arise. They were discovered in mid-nineteenth-century France by a shy mathematical physicist named Émile Mathieu. He was born in 1835 in Metz, a town in north-east France near Luxembourg and the German border. He studied at the École Polytechnique (the university Galois failed to get into), completed the course in 18 months, and went on to write a PhD thesis about multiple transitivity. This led to his five gems, which are famous in pure mathematics, though Mathieu's claim to fame during his lifetime was as a mathematical physicist. Fame is perhaps the wrong word for this quiet man with his shy and retiring nature, though he was well known to his colleagues in mathematical physics, and highly respected. His career took him to a chair at Nancy, just 30 miles south of his birthplace, and he lived there until his death in 1890.

Mathieu published his findings in 1861. He had discovered two groups of permutations that were 5-fold transitive, one permuting 12 symbols, the other permuting 24. They are now called $M12$ and $M24$. Unfortunately, Mathieu didn't manage to convince everyone that these groups actually existed. He certainly felt he had demonstrated the existence of $M12$, and at the end of his paper he writes that, 'by similar methods I have found a 5-fold transitive group permuting 24 symbols'. In 1873 he published a further paper on $M24$ in which he promised to live up to his earlier claim, suggesting that there had been some doubting Thomases. This still didn't satisfy all critics, and an American, G. A. Miller, even cast his doubts into the title of a research paper, 'On the purported 5-fold transitive groups of É. Mathieu'. Miller computed numbers with increasing incredulity, until he hit a contradiction. But arithmetical errors on his part rendered the paper useless. Apparently he realized this later, and in his collected works this paper appears without comment. It took a long time before all doubts were finally overcome. Finally at a Hamburg seminar in 1934–5, Ernst Witt gave a definitive clarification, leading to a remarkable design on 24 symbols, whose symmetry group was $M24$. That proved $M24$ existed to everyone's satisfaction.

Witt studied in the wonderful mathematics department at Göttingen, where the great David Hilbert had insisted on a permanent job for Emmy Noether, saying that a university was not a bathing establishment. She was Witt's thesis supervisor, and his PhD was in 1933, the year the Nazi party came to power. Emmy Noether lost her job, since she was Jewish, and Witt did the unthinkable in joining the Nazi party and the SA on 1 May. Noether's seminar took place in her home, and Witt turned up one day in his SA uniform. This makes him appear beyond the pale, but comments from those who knew him later suggest that Witt was rather naive. He seems to have had no interest at all in politics, nor to agree with the anti-Jewish attitudes of the Nazis, and apparently he later wanted to get out, begging Reinhold Baer to find him a position in Manchester. Baer had also been a PhD student in Göttingen at one time, and then found a position in Jena. Being Jewish he abandoned Germany when the Nazis took over, though he returned in 1956 – we shall hear more of Baer later. Witt stayed in Germany, and after the war he took a position at the University of Hamburg. This city lay in the British sector of Germany and the British military government dismissed him from his position. His accounts were blocked, he was forbidden to enter the university, and his food ration cards were withdrawn. He appealed the dismissal and several German mathematicians came to his defence, confirming that he had not been politically active, a fact confirmed by the Nazis' own negative assessment of his commitment after a compulsory National Socialist course for lecturers during August 1937. He was reinstated and served as a professor of mathematics in Hamburg until his retirement in 1979. He remained rather honest and naive, never quite realizing how shocked people could be by his flirtation with the Nazi party. For example, in the academic year 1960–1 he went to the Institute for Advanced Study in Princeton, and I leave it to one of his later students, Ina Kersten, to explain what happened. 'There, one day, during a discussion about a member of the National Socialist Party, he felt obliged to declare that he had also been a member of the party. To behave otherwise would have seemed insincere to him. He found, to his utter astonishment, that his contacts with his colleagues were suddenly severed.'*

Witt was a fine mathematician and fits the picture of an academic solely interested in his own subject. He felt mathematics should be taught in a unified way rather than split into separate disciplines such as algebra, geometry, and so on, and redefined the mathematics courses in his own department.

Witt's design for Mathieu's largest group $M24$ is similar to the one I showed earlier with seven symbols and blocks having three symbols each. In that case each pair of symbols lay in exactly one block, and the group of symmetries was 2-fold transitive on the seven symbols. In Witt's design there are 24 symbols, and the blocks have size 8: each set of five symbols lies in exactly one block, and the group of symmetries is 5-fold transitive. There are 759 blocks, a number explained in Appendix 2. This design is very exceptional, and very important. It is the first milestone on the road to the Monster, and we shall come back to it later.

In total, Mathieu found five exceptional symmetry atoms: $M11$, $M12$, $M22$, $M23$, and $M24$ (the number indicates the number of symbols being permuted). When I was a graduate student I wanted to understand these groups of permutations, and knowing you should walk before you run, I started with $M11$, moved upwards to $M12$, and then to the larger Mathieu groups. This was how Mathieu originally discovered his groups, but once $M24$ is known it is easier to start there and work downwards. At any event, when I reached $M22$ a question occurred to me, and at teatime I decided to ask an expert. This was in the Mathematical

Institute at Oxford, where teatime was a well-established tradition, and you could guarantee to see people on a regular basis. As soon as I formulated my rather technical question, and put it to Peter Cameron, he said 'It doesn't exist.' I was a bit taken aback, and must have looked puzzled. He went on to explain that $M11$ was not a sub-group of $M22$, as I had obviously assumed, and if it were it would lead to a new symmetry atom. He had already looked for it, and it wasn't there.

As a mere student working in a slightly different area of mathematics, I simply didn't know enough, but expertise like this is only attained after learning and practising on a daily basis, and it can be lost if no young people come along to learn from the masters. We shall come back to this point in connection with the discovery and classification of all finite symmetry atoms, where the techniques were so formidable it was feared future generations might be unable to understand them. They would then be lost, like the knowledge of Egyptian hieroglyphs,* but let us get back to our story.

You may remember that we left Janko working away at an imaginary symmetry atom, trying to find a contradiction. The more he worked – finding contradictions that melted away on further investigation – the more he began to think there was really something there. And the longer he spent on it, the better he understood it, so he gave up on the contradictions, assumed the imaginary object was real, and set about working out the details.

The way to establish the existence of a symmetry atom, known only through its cross-sections, is first to work out its character table. A character table is a very efficient way of writing down information about a group. It is a square array of numbers that gives an immense amount of useful information. For instance, it is very useful in finding sub-groups of the group in question, and you figure it out in pieces, just like a jigsaw puzzle. The more pieces you have in place, the easier it is to extend to a larger portion of the table.

Once Janko had figured out the whole character table he was able to establish that this strange symmetry atom, if it existed at all, had to be able to operate in seven dimensions using 11-cyclic arithmetic. If this sounds weird, it is. But Janko defined a couple of symmetry operations in seven dimensions, which together would generate the whole thing, just as we can generate all the zillion symmetries of the Rubik cube by 90° turns on the various faces. Using computers it was soon shown that Janko's two symmetry operations did the trick, and the new symmetry atom was born.

During this time, Janko stayed in touch with Thompson, and when the new symmetry atom began to emerge, Thompson wrote asking what had happened to the last contradiction. Janko wrote back: 'I never did find my mistake.' As Thompson said, 'That shows how tricky these [exceptional symmetry atoms] are. You keep killing things off left, right and centre, but they're still there.' In the paper Janko submitted in 1965 (published in 1966), he called this new symmetry atom J. It is now known as $J1$, because Janko made further discoveries, as we shall see.

At the time of its discovery, some people expected that $J1$ would be better understood when we grasped the geometric pattern that it was preserving in seven dimensions, but this hasn't happened because the pattern is rather unnatural. Daniel Gorenstein, who later coordinated the great 'Classification' project – the immense work showing that apart from a limited number of exceptions we already had a complete list of all finite symmetry atoms – wrote in 1982, 'there is no natural geometry associated with it. . . . Thus no pat reason for the existence of this [symmetry atom] has been found . . . [it] could have been discovered only in the process of treating some general classification problem.'* If Janko hadn't discovered it, someone else would have found it, but only by a great deal of hard work. For example, Marshall Hall Jr at the California Institute of Technology systematically determined all symmetry atoms of size less than one million – in other words, having less than one million operations – and would have hit $J1$ in the process because its size is 175,560. This number, by the way, is *small* for an exceptional symmetry atom. The smallest is $M11$ at 7,920, followed by $M12$ at 95,040; $J1$ is the third smallest.

Janko's discovery really set the cat among the pigeons. No longer could one assume that the periodic table, along with the multiply transitive groups of Mathieu, was a complete list of symmetry atoms. And if one was missing, how many more were out there?

Janko himself, once he had found $J1$, immediately set about looking for others. He tried lots of cases, but if they didn't produce anything interesting he moved on, and didn't waste time writing up failures. This was a treasure hunt and once he had looked in the likely nooks and crannies of one room he moved to the next one. If other people wanted to search in the same places he had looked, that was their business, but he had a good nose for locating these things and soon found where another one might be hiding. One of the Mathieu groups had a cross-section that intrigued him because he could extend it using the cross-section in

$J1$. This bigger cross-section did not occur in any symmetry atom known at the time, but Janko had really hit the spot. It yielded not one new symmetry atom, but two!

At first it appeared there was just one, and Janko worked out its size to be 50,232,960. All its cross-sections had the same shape, but there was an outside possibility of another example with two different cross-sections. Janko was at Monash University (near Melbourne) at the time, and a German mathematician named Dieter Held was there too. Held recalls that:

> Janko told me there was this other case ... with precisely two cross-sections. He offered to sell it me for ten Australian dollars, but I declined as I – and also Janko – did not believe that it would be fertile. But the next day Janko came up with size 604,800 for the second case. Existence had not been proved for either of these two possibilities as it was the case with $J1$, but it was pretty clear that Janko had discovered his second and third groups.

Of course, Held didn't really believe Janko would give up all work on the second case for ten dollars, so he didn't really see it as a lost opportunity. And a couple of years later, Held had the good fortune to find a new symmetry atom of his own, again by using the cross-section method.

Janko had now produced strong evidence for two new symmetry atoms, and they were later named $J2$ and $J3$ in his honour, but in the opposite order – $J2$ being the smaller and $J3$ the larger of the two. By the time his paper appeared, $J2$ had already been constructed, as a group of permutations on 100 symbols – more on that later – but $J3$ was not easy to construct, and needs at least 6,156 symbols. It is therefore rather odd that in his paper about them in 1968, he starts with the following sentence: 'By studying the structure of the five [symmetry atoms] of Mathieu one observes at once a slight gap in the list of known [symmetry atoms].' Perhaps Janko wanted to portray his discoveries as humble observations that anyone could make, but on the other hand he called his first group *J*, after his own name, so he must have been quite proud of his work, and rightly so.

Janko hailed originally from Zagreb in Croatia, but being politically suspect he was unable to get a university position, and became a high school teacher in Mostar (in Bosnia). At that time in the late 1950s, Croatia and Bosnia were part of Yugoslavia, which was under Communist rule, and anyone who didn't toe the line was politically suspect. Fortunately for mathematics, he won a fellowship to Germany in the early 1960s, and he and Held originally met in Frankfurt. Janko didn't return to Yugoslavia, his passport became invalid, and he needed to find an academic position somewhere. He thought of trying Canada, but the Canadians required an exam in English, so he went to Australia where there was no such requirement. After a year in Canberra he found a permanent position in Australia, later moved to the USA, and finally settled in Germany. His new symmetry atoms – and he discovered strong evidence for a fourth some years later – demonstrated a genius for finding these things, and won him a full professorship at the University of Heidelberg in Germany.

The existence of Janko's new group *J*2 was quickly established after he found the evidence.* Two constructions by hand were given independently by Jacques Tits and Marshall Hall. Both created it as a group of permutations on 100 symbols, Tits in a geometric way, and Hall in a more group theoretical way. Hall gave a talk on his construction at a conference in 1967 at Oxford. As a group of permutations on 100 symbols $J2$ was as close to being 2-fold transitive as you can get without stepping over the line, and two members of the audience were excited about this because it reminded them of a situation they understood rather well. Donald Higman from the University of Michigan, and Charles Sims from Rutgers, wondered whether a similar trick would work in similar circumstances that they happened to know about. They started working out some details, and continued puzzling over it for the next day or two.

The last day of the conference was Saturday 2 September, and it ended with a conference dinner. Sims recalls that 'After the main part of the meal, we were all asked to leave the hall while the staff cleared the tables and prepared for dessert and coffee. As Don and I walked around the courtyard of the college, we again talked about it.' While walking together they did a computation, and it yielded numbers that fitted together well. 'We were sure we were on to something, but at this point it was time to go back in for dessert. After the dinner we went back to Don's room to continue.'* Working together with paper and pencil, they simply carried on until they finally got it, and in the early hours of Sunday 3 September 1967 they had a new symmetry atom. This was extraordinary. Some people laboured for years to discover these things, yet Higman and Sims found theirs in a space of less than 48 hours.

I was a graduate student at Oxford about ten years after that conference, and remember once walking down the street towards the Mathematical Institute. Ahead of me was a grey-haired older man I had never seen before, walking slowly with a stoop. I was about to overtake him as I reached the door of the building, and was faced with the dilemma of either going in front of him, which would be a bit rude, or behind him, which would be a bit slow. To my astonishment he turned left and entered the door himself. I wondered whether to point out that this was the Mathematical Institute and one couldn't just walk in off the street without a good reason. But that wasn't really my business so I kept quiet, and went to get a cup of tea and some biscuits. It was that time of the afternoon.

He did the same. And then sat down opposite me. At this point I felt politeness dictated we communicate, so I asked him if he was perhaps visiting from some other university. Not exactly, he said, because he was now retired and was visiting England from Southern California. I felt it would be polite to ask his name, and he said he was Marshall Hall. 'Oh my goodness,' I said, 'we used your book for our undergraduate course, though we only covered a small part of it.' And then I remembered hearing that Marshall Hall collected coins, so I asked if that was true. Yes it was. I explained that I too had collected coins as a teenager and had been particularly keen on early English ones. What sort did he collect?

In response, the old gentleman fished inside his sports jacket and produced a polythene folder with pockets, each of which contained a coin. They were gold, they were ancient, with Greek letters, and in perfect condition. I'd never seen anything like them. Here was an elderly mathematician walking the streets carrying mint-condition gold coins from ancient Greece worth enough to buy a smart Porsche, if not a house. And who would know? I didn't dare ask whether he was worried about walking along the street carrying such valuables, but on thinking about it afterwards, I could see it was far preferable to leaving them in the hotel. A dealer from London was once visiting the coin fair in Chicago and when he was leaving the city from O'Hare Airport he placed his bag of rare coins in the metal detector. He never saw them again.

When Hall gave his talk at Oxford ten years earlier, and it inspired Higman and Sims to produce another exception, these new symmetry atoms seemed to be two a penny. First Janko had found one, then he looked for another and found two. Then Higman and Sims found one, and others soon followed, either by using the cross-section method or by studying groups of permutations. These things seemed to be popping up like goblins at a fairground. Pandora's box had been opened and soon another surprise was in store, using multidimensional geometry. A remarkable structure in 24 dimensions had recently been discovered, though not yet fully investigated. It yielded further exceptions, though the reason for its creation was not to find new symmetry atoms but to solve a problem in radio broadcasting.

12

The Leech Lattice

Wherever groups disclosed themselves, or could be introduced, simplicity crystallized out of comparative chaos.

E. T. Bell, *Mathematics, Queen and Servant of Science*

In the early days of radio broadcasting, reception was often dis-
turbed by background noise and distortion. You could be sitting
listening to music, but extraneous noises would disturb the sound,
even though the highest standards of acoustic cleanliness were
being maintained in the broadcasting studio. Looking for a way
of alleviating the problem, Claude Shannon at Bell Labs proposed a solution in the 1950s. His idea was to send a radio signal
as a series of very short bleeps; faulty reception in each bleep
could be automatically corrected, cutting down the distortion.
His method associated each bleep to a point in a lattice, and
transmitted the coordinates of that point. Distortion would move
the points slightly off the lattice, and could be corrected by moving
each point back to the nearest lattice point. For this to work well
he needed lattices in multidimensional space, and mathematicians
started looking for them.

Why multidimensional space? The reason is that we need to
keep the points reasonably far apart, so that a small level of distortion doesn't shift one lattice point too close to another, but at

the same time we are trying to pack in as many points as possible. The more dimensions we have the more points we can pack in. To see why this is the case, think of each point being in the middle of a box that protects it from other points, keeping them all a suit-able distance away. Then think of arranging a million boxes. If you put them in a line it will be a million boxes long. If you arrange them in a square each side will have a thousand boxes. If you arrange them in three dimensions the sides will be only a hundred boxes long. And if you could arrange them in six dimensions the sides would be only ten boxes long. The more dimensions there are, the more points you can have within a given diameter, yet keep them reasonably far apart.

The number of dimensions is not the only issue – the arrange-ment of the points is important too. Here are two different arrangements of points in the plane.

The tighter arrangement is on the right. Both have the same mini-mum distance between points, but the one on the right has more points in a given area. To see this, think of surrounding each point by a circle. If we want a minimum distance of one centimetre, then we take the radius of each circle to be one-half centimetre. When two points are exactly one centimetre apart, the circles will touch. When they are more than one centimetre apart, the circles will be disjoint from one another.

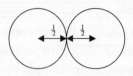

Packing points in the plane so that any two of them are at least one centimetre apart is exactly the same as packing circles of radius one-half centimetre so that no two circles intersect. To see the best way of doing this, think of arranging same-sized coins on a table. The lattices of points shown above give the following two arrangements of circles.

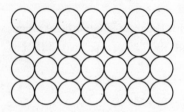 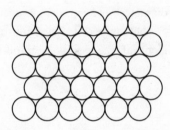

The arrangement on the right is tighter, a fact we can quantify by comparing the space within the circles as a percentage of the total space available. This is called the packing density: in the arrangement on the left it is a little under 79 per cent, but on the right it is about 90 per cent. Another difference is that in the arrangement on the left each circle touches only four neighbours, but on the right each circle touches six neighbours. The more circles that touch one another, the greater the packing density, and in two dimensions one circle can touch a maximum of six others.

Having tackled the problem in two dimensions, let's ask about three dimensions. What is the best way of packing balls, all the same size? And how many balls can touch a given ball? Mathematicians have calculated the answer to these questions. For the best packing, you cannot do better than you often see in a pile of oranges arranged at a fruit stand. First arrange one layer of oranges, like coins on a tabletop. In the picture below, they are shown as circles with their centres at the points marked *a*.

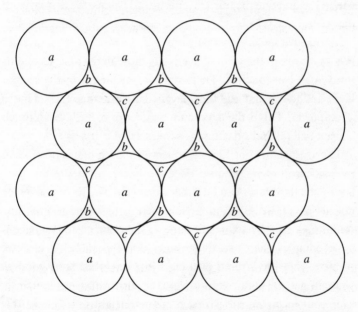

Then pack the second layer in a similar way, but avoid having one orange immediately above another. It is better to have the oranges in the second layer nestle into the spaces so that each one touches three oranges in the first layer. There are two ways to do this. The oranges in the second layer can be placed with their centres above either the points marked *b*, or those marked *c*. You cannot do both at the same time because a point marked *b* and a point marked *c* get too close.

When we pack the third layer there are again two choices. If we packed the second layer with their centres above the points marked *b*, then we must pack the third layer with their centres either above the points marked *a* (that is, directly above the first layer), or above the points marked *c*. These two choices are different: in the first one the third layer is directly above the first layer, in the second one it isn't.

In these packings, how many oranges can touch a given orange? For an orange in the second layer, only those in the first, second or third layer can touch it. There are six touching it in the second layer, and in the first and third layers there are three each. That is a total of 12. This is the maximum number of balls that can touch a given ball (all balls having the same size, of course).

The arrangement of 12 balls touching a given ball is possible in two different ways, because once we have settled the first two layers, the third layer can be either directly above the first layer, or not, as we please. By comparison there is only one way of surrounding a coin by six neighbours, so three-dimensional space is more complex than two dimensions, and this makes it difficult to prove that the layered packing I just described is as good as you can get. It might seem obvious to anyone but a mathematician, but in mathematics one needs watertight proofs, and this became a famous unsolved problem called the Kepler conjecture. Although Kepler posited the conjecture in 1611, it resisted all attempts at proof until 1998, when Thomas Hales at the University of Pittsburgh finally solved it. His method used computers by first reducing what was a problem about an infinite number of things (the Kepler conjecture considers an infinite number of spheres in an infinitely large space) to a very large, but finite, number of problems. Each problem involved a finite structure that he compares to a sculpture made of cables and struts. There were roughly 100,000 such structures, and by using a computer to analyse them all he was able to complete his proof of the conjecture. Checking the work took years, and the paper only appeared in 2005.

The Kepler conjecture made no assumption about the centres of the spheres being the points of a lattice. If you assumed this, there was a relatively simple proof, but the assumption was not valid, because there are layered packings as tight as any lattice packing but without conforming to a lattice. However, this need not bother us because we are concerned with symmetry, and can stick to lattices.

We arrived at the subject of lattices from Shannon's idea for reducing distortion in radio broadcasts, and this means finding good lattices in more than three dimensions. How do we do this? In dimensions 4, 5, 6, 7, and 8 the higher dimensional crystals from Chapter 9 can be used, particularly the exceptional ones of type E. They form the basis for some excellent lattices, but in dimensions higher than eight the exceptional crystals disappear, and really tight lattices get more difficult to find. However, in dimension 24 something very extraordinary appears. It was discovered by a man named John Leech, and is called the Leech Lattice.

John Leech was a mathematician who became very interested in computing in its early days. He worked for some years in industry, then in the Computing Laboratory at Glasgow University, and eventually became head of computer science at the University of Stirling in Scotland. In the early 1960s he had the excellent idea of using Witt's design. This was a design on 24 symbols that Witt used to construct Mathieu's groups (page 133), and Leech now used it to construct a lattice in 24 dimensions. He published his first paper on this in 1964, and then in 1967 added extra points to give a yet tighter packing, now known as the Leech Lattice. It cannot be bettered. It is the tightest possible lattice packing in 24 dimensions, though a proof of this fact was only announced in 2004.*

In the Leech Lattice, each 24-dimensional sphere touches 196,560 others. This number is explained in Appendix 3, and will appear later in connection with the Monster, though when Leech published his lattice the Monster was nowhere in sight. But the strange new symmetry atoms of Chapter 11 were being discovered, and Leech was intrigued.

His construction of the lattice showed that it admitted a lot of mirror symmetries, along with Mathieu's largest group of permutations, and he wondered whether there wasn't more. He had a feeling there was, and that an enormous new symmetry atom would emerge. Since his lattice was so exceptional, this would surely be a symmetry atom not in the table. Leech tried getting group theorists interested in his lattice, and as he said himself, 'I dangled the problem under various noses . . . but Conway was the first to take the bait.'

John Horton Conway was born in Liverpool in 1937, and went to Cambridge as an undergraduate when he was 18. He claims to have been painfully shy at one time but, 'When I was on the train from Liverpool to Cambridge to become a student, it occurred to me that no one at Cambridge knew I was painfully shy, so I could become an extrovert instead of an introvert.' Conway became an outgoing and engaging character but, like many other creative people, was not always a success as a student. He preferred to work on what interested him rather than on what was in the syllabus. He loved games and started inventing his own, but it did nothing to help his exam performance, and it wasn't clear he could stay on and take a PhD. Fortunately he did, but as his graduate studies were coming to an end, he needed to find some sort of position.

The department chairman, Professor Cassels, asked him what he had done in applying for a job, and said 'We're advertising a position. You should apply.' 'How do I do that?' Conway asked, so Cassels took out a piece of paper and, sitting on a wall outside King's College, wrote, Dear Professor Cassels, I would like to apply for . . .'. Conway didn't get the job, but when a similar position came up the next year, the head said, 'Unless I hear from you, I'll take your letter as an application for next year.' This time he was lucky, and remained at Cambridge until he was lured away by Princeton over 20 years later in 1986.

Conway's first big break was his work on Leech's lattice. In autumn 1967, John Leech went to Harwell to spend a year at the Atlas Laboratory. This was Britain's large computer centre, located near Oxford. John McKay, he of the later Moonshine speculation, was there too, and the pair of them went to mathematics seminars at Oxford University together. Graham Higman* at Oxford had been working on Janko's third group $J3$, and McKay got involved in using Higman's results to construct it as a group of permutations. Leech tried to interest Higman in his lattice, and McKay tried it out on the people at Cambridge. He went there to talk to Thompson and others about constructing $J3$, and told Thompson and Conway about the lattice, but nothing happened immediately. The trouble was that since the recent discoveries of new symmetry atoms, Thompson had heard heaps of suggestions of where new ones might be lurking, and most ideas led nowhere.

For Conway it was a different matter. He was not really a card-carrying group theorist, and when he got his first university appointment at Cambridge in 1962 he was working on mathematical logic and finite mathematics. Things were not going well, and he wrote later that 'I became very depressed. I felt that I wasn't doing real mathematics; I hadn't published and I was feeling very guilty because of that.'* Conway was intrigued by the Leech Lattice, and took a look at the first paper by Leech. He then phoned Leech, and Leech said look at the later paper, which had just appeared in print. When he did, he agreed there ought to be a large group of symmetries, and tried to persuade Thompson to take an interest. Thompson declined, but said that if Conway could work out the size of the group of symmetries, then he would believe there was something in it.

This looked like very hard work. Conway had a young family with four daughters, and was doing extra teaching to bring in the bread. How on earth was he to find the time to work on this interesting and difficult problem? He waited until the summer vacation and discussed it with his wife. 'This could make my name,' he assured her, and they agreed that he would get two uninterrupted periods a week to work on it. One was to be Wednesdays from 6 p.m. to midnight – the other was Saturdays from noon to midnight. Before I recount how his first day's work went, let me explain what the problem was.

If you fixed a point in the Leech Lattice, the neighbouring points were obtained in three separate sets. In Leech's original paper there were two sets, one giving 97,152 points, and the other giving 1,104. The second paper added a third set of 98,304 points. This gave a grand total of $97,152 + 1,104 + 98,304 = 196,560$ points neighbouring a given point. There were plenty of symmetries that kept these three sets of points separate, but what Conway needed was a symmetry that mingled them, allowing a point in one set to go to another set. Physicists have a similar problem with elementary particles. They, the particles, are in different families, and physicists would like to see particles in one family transform to those in another. How can this be achieved, and what new symmetries will reveal themselves in this way? Conway's first job was to see how likely this was in the case of the Leech Lattice. If you take one point and a neighbouring point, how many points are neighbours of both? If you get the same answer for any pair of neighbours, this is evidence that one pair of neighbours is equivalent to another. You can then add a fourth point neighbouring three mutual neighbours, and do a similar calculation. Conway did this and things worked out well. The evidence piled up, and he was convinced there was a large group of symmetries. Now he really wanted to work out the size of this group.

At noon on Saturday, as agreed with his wife, he started work, 'I had a last cup of coffee, kissed the wife and kids goodbye, locked myself in the front room, and started to work.' With a 12-hour period ahead of him he took a long sheet of white paper and wrote down everything he knew about the Leech Lattice. By 6 p.m. he calculated that it should yield a symmetry atom of the following size, or possibly half this:

$$2^{22} \times 3^9 \times 5^4 \times 7^2 \times 11 \times 13 \times 23 = 8,315,553,613,086,720,000$$

He phoned Thompson.

At this point it may seem that Conway and Thompson, being colleagues in the same department, would be on a roughly equal footing with one another. This is not the case at all. Conway was a junior faculty member with little serious mathematical work to his name, and his technical mastery of group theory was relatively poor, while Thompson was, well, stratospheric. As Conway says himself, 'I was a bit in awe of him actually, because he was the best group theorist in the world, and everybody knew that. And I thought he was a very serious person.'

Later, when Conway had analysed the symmetry of the Leech Lattice in great detail, he was invited to give talks all over the place. One of the first was at Oxford, and at the end of the talk a graduate student asked, 'How do you know your new group is simple?' In other words, how do you know it cannot be deconstructed into anything simpler – or to use the term in this book, how do you know it is an *atom* of symmetry? Conway was slightly taken aback because he didn't really have an argument, so a faculty member at Oxford named Peter Neumann answered the question by putting a simple argument up on the blackboard. 'I felt like a fraud in all these talks', said Conway. However, Peter Neumann was very impressed with the talk and solicited a paper for the *Bulletin of the London Mathematical Society*, promising swift publication. Conway wrote the paper that autumn, and it appeared soon after, in 1969.

Conway may have felt a lack of technical expertise at that time, but what makes a young mathematician look good is independence of spirit and creativity. Technical mastery will be acquired by learning from those who already have it, but creativity doesn't work this way. Mathematics has its fair share of brilliant young people who seem to be able to learn anything with great rapidity.

Some of these people can assimilate technical material at a tremendous rate, but never go anywhere because they have no creative ideas of their own. Conway had independence and creativity in buckets, so he really had nothing to fear.

When Conway phoned Thompson giving him the number $2^{22} \times 3^9 \times 5^4 \times 7^2 \times 11 \times 13 \times 23$, saying either this number or half of it was the size of an atom of symmetry, Thompson phoned back 20 minutes later, told him he needed to halve it, and that there were two other new symmetry atoms associated with it. Conway recalls that 'We used to joke that if you wanted a new symmetry atom all you had to do was work out its size, pick up the phone and dial John Thompson, and dictate the number. The results could be quite spectacular.'

But there was still a big problem. Conway had worked out the size, Thompson had confirmed it made sense, but did the new symmetry atom really exist? Conway was working in 24 dimensions, and he needed one new symmetry that wasn't visible using Mathieu's group, along with a group of mirror symmetries that he already knew about. The mirror symmetries were generated by 16-dimensional mirrors, all being permuted by Mathieu's group, and one further symmetry should generate the whole of this new group. In order to write down a symmetry in 24 dimensions you take 24 axes, and specify where each one goes to. This involves writing down the coordinates of 24 points. Each of these points has 24 coordinates of its own, so you write down 24 sets of 24 numbers, in the form of a matrix. This is what Conway did, 'filling in the entries piece by piece'.

It was hard work because the matrix had 576 entries and he couldn't afford a single mistake. He finally completed it, and although he wasn't quite sure this matrix would do the trick, he was ready to call it a day.

> Anyway I telephoned Thompson again and told him that I had this matrix, but that I was feeling exhausted, even though it was only ten o'clock, and was going to bed. I would talk more about it tomorrow. Then I hung the telephone up.
>
> Then I thought, 'No, I won't. I'll just see if I can at least see, in principle, how to do it.' . . . Anyway it suddenly dawned on me as soon as I had finished telephoning him the second time that I was being stupid.

Conway suddenly had an idea of how to test the matrix. It involved doing 40 calculations. He did one in detail, and the result was fine. If he just did 39 more calculations like this it would settle the matter. But by this time he was very tired, and said to himself, 'It's all going to work and so, really now, I'm going to bed.' Going to bed when something exciting like this is left unfinished is a bit unsatisfactory, so Conway stayed up for a while.

> I just said, 'Well, how bloody stupid to give up,' and so I carried on. At a quarter past midnight, I telephoned Thompson again, saying it was all done. The group was there. It was absolutely fantastic – twelve hours had changed my life. Especially since I had envisioned it going on for months – every three days spending six or twelve hours on the damn thing.

What Conway had shown in a remarkable 12½-hour period was that the symmetry group of Leech's Lattice was larger and more complex than had hitherto been realized and, as he said later, 'Those 12½ hours were the most important of my life.'

The next day was Sunday, and Conway and Thompson met in the

mathematics department. They worked all day on the new group and their discussions continued for the whole week. 'I got a fantastic education from Thompson', said Conway. Among the first things to come out were two further new symmetry atoms, making three in total, and in honour of Conway they are referred to as *Co*1, *Co*2, and *Co*3, though Conway himself named them •1, •2 and •3 (pronounced dot-one, dot-two and dot-three). The group •1 was obtained by fixing one point of the Leech Lattice. If you fixed two points as close as possible to one another, you got the group •2. If you fixed two points at the next possible distance from one another, you got the group •3.

These groups were not all. Fixing two points at other distances yielded two other exceptional symmetry atoms, which in Conway's notation were •5 and •7. The second of these was identical to the Higman–Sims group, discovered six months earlier at a conference in Oxford, and •5 was the same as a new group of permutations only just discovered by Jack McLaughlin at the University of Michigan. This was exciting stuff. McLaughlin's group wasn't yet published and Conway hadn't even heard of it. But Thompson had, and Conway recalls that 'This was one of the things that really convinced him.'

Now that Thompson was sure the new groups were genuine, he examined them in some detail, and pulled two more exceptional symmetry atoms out – one was Janko's second group *J*2, and the other an exceptional group of permutations discovered in a different way by Suzuki, he of the Suzuki family of groups. If Conway had been able to investigate the Leech Lattice a year or two earlier he would have turned up seven new symmetry atoms instead of three!

In total Leech's Lattice yielded 12 monkeys: five Mathieu groups, three Conway groups, and four others. These new discoveries changed Conway's life. 'I had always felt guilty, felt that I was not good enough, but the new discovery in 1968 released me from worry, and led me to do some really good stuff.'

Conway continued to be interested in games, both competitive and solitary. He has written two books on the subject,* and his 'game of life' – which isn't a game in the usual sense, but a way of producing interesting patterns, with interesting implications philosophically – has been featured on television programmes and can be found easily on the Internet.

Conway remains the world's top mathematical gamesman, and a genius at finding clever notation and methods to study complicated phenomena. In his earlier years he used to look rather eccentric, and was seen around Cambridge University in open sandals, even in winter. At a conference at McGill University in Canada back in the 1970s when there was 18 inches of snow outside, his sandals were thoroughly soaked, so he took them off when he arrived to give his lecture, and stood in bare feet. The mathematician who was introducing him to the audience had prepared a little rhyme.

> Let's make a toast to Conway, John
> The celebrated Cambridge don
> You'll know him by his unshorn locks,
> And by his frequent lack of socks.*

The remarkable Conway will appear again before the end of this book.

13

Fischer's Monsters

In great mathematics there is a very high degree of
unexpectedness, combined with inevitability and economy.
G. H. Hardy, *A Mathematician's Apology*

In any creative activity it can sometimes be very useful to go back
to fundamentals. The Italian Renaissance, for example, harked
back to the ideals of classical art and architecture. Mathematics is
like this too, and some wonderful advances are made by going
back to basic questions.

Bernd Fischer did what many excellent mathematicians have
done before him and will continue to do in the future. He went
back to a simple sounding problem. Suppose you have a group of
permutations. The simplest permutation is one that switches the
position of two objects, leaving everything else alone. It is called
a transposition. We met transpositions in Chapter 3 when we
considered even and odd permutations. As abstract operations
they have order 2 – if you do the same transposition twice then
everything is back to where it was – and Fischer asked a simple
question: which groups of permutations are generated by operations of order 2 that behave like transpositions? He did not need
the operations to be transpositions in the usual sense, but merely
to behave like them in a way I will make precise later, and he

concentrated his attention on symmetry atoms, or at least groups that are very close to being symmetry atoms. This led him to three big surprises, and they in turn took him on a slightly different track that led eventually to the Monster, though he certainly didn't realize that at the time.

Mathematics was Fischer's abiding interest since childhood, when he loved to do homework for older boys at school. He was fortunate later in having a stimulating teacher:

> In high school I had a very good mathematics teacher. He had been an assistant for three years before the war in Darmstadt, where he worked on rocket trajectories. This was sophisticated mathematics – it used differential equations, and had to take into account the change of air pressure with altitude. He was a fantastic mathematics teacher, and when I went to Frankfurt University I had no need to attend any of the lectures on differential equations.

Germany was a leader in rocket science, producing the V2 rocket during the later stages of the Second World War. Fischer's teacher was not a rocket scientist but a mathematician, and his way of using mathematics to deal with physical problems inspired Fischer, who went to university intending to take a master's degree in physics, and then do a PhD in mathematics. But at university he encountered a professor named Reinhold Baer, who had returned to Germany from the USA. 'From the beginning I was fascinated by the way he did mathematics, and the way he spoke to students of my age. I took analysis from him and he would make mistakes on purpose to get the students to correct him.' Baer's influence turned Fischer from a would-be applied mathematician to a pure one. He admired Baer's attitude: 'He was very broad, and wanted to have all parts of mathematics in his seminar. He was an excellent senior professor. You could work on anything you wanted, which for German universities at that time was very unusual.' Baer provided a wonderfully stimulating environment for creative work. 'He got an extraordinary group of students and visitors together.' Some visitors came for extended periods, others just to give a talk. 'Almost everyone in this area of mathematics came to Frankfurt. Tits was a frequent visitor, Thompson came, Janko came; there were very few exceptions.'

Before the war, in the spring of 1933, Baer took a long vacation in Italy with his wife and son. When Hitler came to power, Baer, being Jewish, decided not to go back home, but went to England without touching Germany, and in autumn he lost his position owing to the new Nazi laws. He found a job at Manchester University, and two years later went to the USA, but he never felt America was his home, and in 1956 returned to Germany. Fischer recalls that 'Baer loved the German university system. He really liked the way a university professor in Germany was independent, and he knew so well how the German system had developed in the nineteenth century; it was as if he had invented it himself.'

When Fischer was a student he looked into various parts of mathematics and started developing ideas of his own. 'I used to go into the library, crack open books and start reading them. One of the things I read about was distributive quasi-groups.' These were not groups, but Fischer got interested because 'it was obvious to me that there was a group around'. He was right, and it led him on to the track of groups generated by operations that behaved like transpositions, and later helped in cutting a path to the Baby Monster. But let us first consider ordinary transpositions.

Two transpositions, one following the other, can give two possible results. Consider a collection of people round a table – interchanging two of them and leaving everyone else in place is a transposition. Let's do two of these in turn: for example, first interchange Anthony with Beatrix, then interchange Charles with Diana.

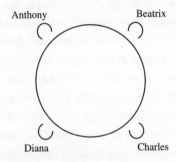

There may be many people at the table, but only four are involved in this transaction. Two separate interchanges take place, and the resulting permutation has order 2; if you do it twice everything is back to its original setting – Anthony and Beatrix end up in their original places, as do Charles and Diana. On the other hand, if both interchanges involve a person in common, then the resulting permutation has order 3. For example if Beatrix interchanges with Anthony, and then with Charles, the result is that all three of them have moved clockwise or counter-clockwise round the table, while everyone else stays in place (with the arrangement in the picture the movement is clockwise). This permutation has order 3 – do it three times in succession and everyone is back where they started.

One transposition followed by another yields either a permutation of order 2, or one of order 3. Now forget about transpositions for a minute, and consider permutations of order 2 with the extra property that one followed by another has either order 2 or order 3. This is what Fischer considered – the permutations he started with didn't need to be transpositions in the usual sense of the term, but he referred to them as transpositions,* and set about finding what groups they could generate.

To cut a long story short – and he didn't do this overnight by any means – Fischer proved a remarkable theorem that said roughly the following. If a symmetry atom, or something very close to it, is generated by Fischer's transpositions, then there are six different cases. It might simply be the group of all permutations on a collection of objects – these groups get big quickly as the number of objects increases, and this is the dull case. The other cases are more interesting: four of them are families of classical symmetry atoms,* and if that were the end of the story, Fischer would have produced a very fine mathematical theorem. But it wasn't the end – the last case was absolutely fascinating.

From Fischer's sixth case emerged three huge symmetry atoms, each related to one of Mathieu's three largest groups of permutations. Mathieu's three largest groups were called $M22$, $M23$, and $M24$, and Fischer's groups became known as $Fi22$, $Fi23$, and $Fi24$. Compared to the size of Mathieu's groups, they are enormous. The first two are symmetry atoms and the third contains a huge symmetry atom of size 1,255,205,709,190,661,721,292,800. This means more than a million million million million symmetries, and outside the periodic table it was the largest discovered so far.

To explain how the Fischer groups are related to the Mathieu groups, think of Fischer's transpositions in terms of mirror symmetries. A transposition interchanges two objects and leaves everything else in place. Think of those two objects as points on opposite sides of a mirror, and the other objects as points in the plane of the mirror. The symmetry across the mirror switches the first two points and fixes the rest. In other words it acts as a transposition.

When we do one mirror symmetry followed by another, the result will depend on the angle between the mirrors. For example, suppose two mirrors are at right angles to one another – think of one mirror interchanging north and south, while leaving east and west fixed, and the other mirror interchanging east and west, while leaving north and south fixed. The combination of the two will switch north with south, and east with west, and the result is a rotation by 180°. This is exactly double the angle between the mirrors, which is 90°, and it is a similar story for any angle. The combination of two mirror symmetries is a rotation, the angle of rotation being twice the angle between the mirrors. If the combination of two mirror symmetries is to have order 2 or order 3, then the angle between the mirrors must be 90° or 60°.

Treating Fischer's transpositions as mirror symmetries is not strictly correct, because I have said nothing about the dimension of the mirror or the space it is in, but the analogy may be helpful. In two dimensions the situation is relatively simple. In three dimensions it is a little more interesting, but Fischer didn't restrict the number of dimensions in any way. If you wish to understand his results from a purely geometric point of view, you may find yourself in some difficulty, because you need lots of dimensions, but Fischer didn't approach it this way. So how did he do it?

One important ingredient, which links up with Mathieu's groups, is the following. Among the zillions of mirrors in the arrangement, Fischer considered a sub-set of mirrors that were all at right angles to one another, and he made this sub-set as large as possible. He then examined the sub-group of symmetries permuting them, and proved it could send any pair to any other pair. He then used this to show that either it would have a straightforward and well-understood structure, or it would be one of Mathieu's three largest groups, $M22$, $M23$, or $M24$. These last possibilities led to the extraordinary sixth case, yielding the Fischer groups, $Fi22$, $Fi23$, and $Fi24$. In his original paper, published in 1971, Fischer called them $M(22)$, $M(23)$, and $M(24)$, and they were sometimes referred to as 'Mathieu groups writ large'. Fischer's use of the letter M was admirably unassuming. He could easily have used the letter F instead.

Fischer's groups are very large, but the way to understand them is to look at the complex arrangement of mirrors. The number of mirrors is far smaller than the size of its group of symmetries. The largest group, $Fi24$, has more than a million million million million symmetries, but the number of mirrors is less than a third of a million – it is 306,936 to be precise. This still sounds large, but we don't need to imagine real mirrors. We can simplify things by using vertices instead of mirrors. Each vertex represents a different mirror, and two vertices are joined by an edge when the two mirrors are at right angles – this creates a network of vertices connected by edges. Dispensing with mirrors and thinking in terms of networks makes it easier for the human mind to comprehend.

For Fischer's largest group, $Fi24$, the network has 306,936 vertices, each one representing a different mirror. In this network, the number of vertices joined to a given vertex, or in other words the number of mirrors at right angles to a given mirror, is 31,671. These mirrors form a sub-network, whose symmetry group is Fischer's smaller group $Fi23$. In this sub-network each vertex is joined to 3,510 others, and the symmetry group on this smaller network of 3,510 vertices is Fischer's group $Fi22$. Within this smaller network each vertex is joined to 693 others, and in this yet smaller network of 693 vertices each vertex is joined to 180 others. The point is that you can build back up again, from smaller networks to larger ones, and you do this by using Mathieu's groups.

This is not supposed to be easy, but it is not impossible either. Mathematicians deal with complex things by building up from simpler components. In some ways it is like creating a design, duplicating it many times and using these duplicate copies as parts of a more complicated design. This is what Fischer did, but when he wanted to move up from $Fi23$ to $Fi24$ he found that there seemed to be two possible extensions, of different sizes. One of these sizes looked absurd: 'It was divisible by a prime number larger than 100,000, which is obviously ridiculous, but you have to rule it out somehow.' Fischer wrote to Feit, who eliminated it using a rather sophisticated technique that had been used earlier by Suzuki, Feit, and Thompson, but Fischer wanted to find his own way to get there, and he did. This was at the end of 1969, and he was now ready to write up a summary of his results.

Fischer's paper appeared in print in 1971; it was labelled part I, and he refers to forthcoming parts II and III that would deal in detail with various cases. They never appeared because Fischer was invited to give a series of lectures at Warwick University in England, and he wrote lecture notes analysing all cases. These notes were freely available, and when I needed a copy some ten years later the mathematics department at Warwick happily obliged. But of course the notes were not available in university libraries, and where another mathematician might have elaborated these notes and published a series of papers, Fischer was off on another track. He communicated his results very readily and inspired others, but wrote very little down for publication. His aim was to do research and communicate it directly with other mathematicians, and he stirred up enormous excitement.

Although most of his work never appeared in print, his notes were widely read and other people published analyses and summaries of his work. Fischer didn't mind at all – he was an energetic man and worked very hard, aided by coffee and a large supply of cigarettes, producing several new symmetry atoms. We now have three of them, with two more to come, the next being the Baby Monster. That didn't appear in print either.

Before we get on to the Baby, it is worth mentioning how Fischer's work inspired others. In his groups generated by mirror symmetries, the angle between the mirrors was either 90° or 60°, which meant that the combination of two mirror symmetries had either order 2 (a half-turn rotation) or order 3 (a one-third turn rotation). A young mathematician in California named Michael Aschbacher followed this up by changing one of the angles between the mirrors. Aschbacher retained the 90° angle (giving a combination of order 2), but changed the 60° angle so that the combination would have order n, where n is an odd number. When n is 3 the angle is 60°, which is Fischer's case, yielding his strange beasts. Aschbacher did the other cases, and published his work in a series of four papers that appeared during 1972 and 1973.

He analysed all symmetry atoms that could arise in this way. It is an intriguing list, but it shows there were no new surprises to be had in this direction. There *was* another big surprise to come – and Fischer found it – but let us first talk about Aschbacher.

The 'Classification project' – the project to find a complete list of symmetry atoms and show that the list is complete – had been kick-started by the great theorem of Feit and Thompson, and then moved forward by Thompson's own work. After him, the second most important contributor was Aschbacher, whose work really took off in the early 1970s. He went straight for the real problems underlying the project, churning out one result after another, and sweeping aside some of the cross-section problems that others had been planning to work on. Some mathematicians who intended making a career of this stuff suddenly had the ground cut from beneath them. As one observer put it, 'they were all going around looking pretty glum because Aschbacher had shown the whole thing was within reach. He took a lot of people who were planning to work on cross-sections, and just zapped them.'

Aschbacher operated at a prodigious rate, but his theorems all had to be refereed before publication. This took quite a toll on the time of other people, and Ernie Shult at Kansas State University recalls that in the mid-1970s he was drowning under them:

> I was supposed to be a referee for a lot of them, but it got to the point where I couldn't be responsible for checking them all. I was getting about six a year, so I didn't have much time for my own research. And some of these papers were about a hundred pages in manuscript. One of them I remember was 120 pages.

These papers were detailed and technical, and writing such a paper takes a great deal of work. Even if you have the outline of a proof in your head, committing it to paper in an organized way is quite a process. If you simply state the main theorem and try writing out a proof, it can be far too cumbersome. Different points in the proof will use similar techniques, so you deal with these techniques separately, splitting them off from the main proof. You write a statement for each one, and then prove the statement. These lesser results are often called lemmas; they may not be of great interest on their own, but are very helpful in proving something more important. All mathematicians use lemmas on the way to a bigger result – they are like pieces of pipe that all have to fit together without leaks. Aschbacher used some of his own lemmas in later papers, not always with identical hypotheses, but although he knew exactly where he was going, you can pity the poor referee who is trying to follow the process.

Aschbacher was writing things so quickly that he usually had no time to rewrite them, and since his proofs were very involved and tersely written, they became very difficult to read. He and Shult even got into correspondence about some of the proofs, because Shult knew that Aschbacher knew that Shult was the referee, so they dispensed with the anonymity and communicated directly rather than go through the editor.

Aschbacher tended to work alone, but other people who later collaborated with him were awed by his enormous grasp of detail, and astonished by work he had done without yet submitting it for publication. For example, Gary Seitz went southwards from Oregon to spend some time with Aschbacher in Pasadena, and said, 'I proposed a question to him and he said he'd already solved it. He opened a desk drawer and there it was! So I tried another question I'd been thinking of and he opened another desk drawer.' We shall hear more of Aschbacher's contributions later.

Let us now get back to Fischer's work. He had considered all symmetry atoms involving a collection of mirrors in which the angle between any two was 90° or 60°. It was natural to try a similar project by allowing 45° angles, and he proposed this problem to a very capable PhD student named Franz Timmesfeld. The problem was diabolically hard, but Timmesfeld took it on, and found that by making one extra assumption he might be able to get a complete solution. In a series of three papers published between 1970 and 1975 he did precisely this. He used geometric methods to study the internal structure of the groups that arise, in the same spirit as Fischer and Aschbacher, and like them he gave a complete list of all symmetry atoms that occurred. Each one appeared in the periodic table. There were no exceptions.

Now Fischer himself re-entered the picture, removing Timmesfeld's extra condition and searching in the depths for whatever was there. He had a good nose for finding exceptions and guessed where another one might be hiding, but he desperately needed time to work on it. 'In summer 1970 I went to Michigan State University and had two months to think about it.' Fischer's new idea involved his group $Fi22$, which could be used to create a set of mirrors in which the angle between any two was 90°, 60°, or 45°. This led him to a larger group in the periodic table that seemed to contain $Fi22$. He lectured on this at Bowdoin College in Maine, and Walter Feit, he of the Feit–Thompson theorem, was in the audience. Feit objected that a result in a paper of Steinberg ruled out this containment of $Fi22$ in the other group, but Fischer knew what he was doing, and didn't accept the objection. As he recalled later, 'Feit said over a hundred people had read this paper of Steinberg. But they communicated my results to Steinberg himself, and two or three days later he said he'd looked over his proof and found an error.'

This type of thing happened more than once in Fischer's work. He would discover evidence for some new group, and get the objection that so-and-so had proved a result ruling out some strange sub-group it would have to contain. But Fischer was always pretty sure of his ground. He was pretty sure the strange sub-group did exist, and an error was then found in the result that eliminated it.

In fact Fischer was on to something very interesting – a symmetry atom that would prove to be very large indeed – but he lacked time to pursue it. After his two months at Michigan State University, he had to return to Bielefeld, where he was to serve as dean, for the coming academic year. 'It was a horrible time to be dean. We had a lot of students coming from Berlin and they knew very well how to disrupt universities. They had learned their tactics from the student unrest in 1968.' It is difficult now to imagine what the problem was and why the students seemed so keen to smash the system, but as Fischer said,

Some of these people wanted to redefine mathematics. For example, I was asked whether a student could write a master's thesis on the merits of Karl Marx as a mathematician. That would be all right, but when he showed me the papers they just involved integration techniques, and went on for pages and pages without any theorems. It was high school stuff, on the level of a good high school student.

Fischer was a good person to deal with idealism and political shenanigans because he is the sort of person who can give a measured response to anything. I cannot imagine him losing his cool over things, but he certainly had a tough job as dean. 'We had meetings starting at 10 a.m. and ending at 9 p.m. I said that for all important things the dean had to be there. They tried to sit me out.' Let us leave Fischer battling with student politics and ideology, and find out what else was going on in the discovery and classification of all finite symmetry atoms. We shall return to him later.

14

The Atlas

A marveilous newtrality have these things mathematicall, and also a strange participation between things supernaturall, immortall, intellectuall, simple and indivisible, and things naturall, mortall, sensible, compounded and divisible.

John Dee (1527–1608), Preface to an edition of Euclid

While Fischer was detained by his administrative responsibilities, other mathematicians were busily trying to show that apart from one or two fissures where strange beasts might be lurking, the symmetry atoms on the list (periodic table plus exceptions) were all that existed. People were proving theorems saying that if a finite symmetry atom had this, that, or the other properties, then it was already on the known list. These results were being proved at a terrific pace, and the whole project – compiling a list of all finite symmetry atoms, and showing that the list was complete – became known as 'the Classification'.

It was a vast undertaking. So many mathematicians were working on different aspects of it that their work was bound to overlap, and at the same time gaps were liable to open up unless all parts were covered. There was clearly a desperate need for someone to orchestrate things and keep track of progress, and as with other aspects of life, someone often rises to the occasion. In this case it was Daniel Gorenstein – he was a larger than life mathematician, a mover and shaker, a fixer, a person who could get things done, encourage people, and oversee the entire enterprise.

Gorenstein dubbed this enterprise 'The Thirty Years War', and it is quite incredible that as the field marshal in charge, he also managed to run the mathematics department at Rutgers University. One of his colleagues said to me a few years later that 'He's the most competent man I ever met. He could do a lot of things at the same time, and do them all well.' Such energy and competence in one person is a rare gift, but 'having a conversation with him was like standing in a gale, and you had to know how to keep your balance'. Gorenstein was a phenomenon, and 'when he was department head and we had meetings, the ideas would tumble over one another on the way out of his mouth. "Would our distinguished department head please finish a sentence", someone once asked. It never happened.'

Here was a man who was firing on all cylinders all the time, handling the vast number of mathematical ideas that went into the Classification, and the mathematicians who were coming up with them. He combined superb organization with tremendous personal drive, and as one of his PhD students remembers, 'When I read my notes of his lectures later, they would come off the page with the sound of his voice, something I never experienced before or since.' Inside the lecture hall his energy and dynamism had 50 minutes to express themselves without serious interruption. Outside the lecture hall it was a different matter, and things were happening at a terrific pace. 'I'd go into his office to spend half an hour with him, and he'd be perpetually interrupted by phone calls. I had the feeling I was at Command Central.'

Gorenstein created a team spirit unheard of in any pure mathematics project at any time, and when he died in 1992 there was a huge outpouring of grief. 'He was like a father to me', said one mathematician to me at the time. Writing later, in 1995, Ron Solomon, a young mathematician who joined the team in the 1970s, says 'Gorenstein . . . provided the optimism, the organization and, in 1972, a "16-step plan" for the completion of the Classification proof.'

This was the high point in Gorenstein's career, which started as an undergraduate at Harvard during the Second World War. One of his teachers was Saunders Mac Lane, who relocated to Chicago in 1947 and was later the thesis advisor to Thompson. After the war, Gorenstein went back to Harvard as a graduate student in algebraic geometry – a different area of mathematics – but in 1957 he got interested in finite groups, and in 1960 Mac Lane welcomed him to the University of Chicago for the big group theory year. That was where Feit and Thompson worked on their great theorem, and Gorenstein got intrigued at the prospect of finding and classifying all finite simple groups – the atoms of symmetry. He started working on the problem himself, partly in collaboration with John Walter at the University of Illinois, and later with others.

By the beginning of the 1970s things were moving fast, and as Ron Solomon writes in 1995, 'The pace of the Classification in the '70s was exhilarating. Not a single leading group theorist besides Gorenstein believed in 1972 that the Classification would be completed this century. By 1976, almost everyone believed that the Classification problem was "busted".'* By the term 'busted', Solomon meant that the basic problem had been cracked and the remaining pieces could now be picked off, albeit with some difficulty and great technical finesse.

173

While Gorenstein himself drew together the strings of this great project, encouraging others to cooperate in the big picture, he freely admits that what made things move so fast was Aschbacher's work:

> It was Aschbacher's entry into the field in the early 1970s that irrevocably altered the simple group landscape. Quickly assuming a leading role in a single-minded pursuit of the full classification theorem, he was able to carry the entire 'team' along with him over the following decade until the proof was complete.*

Marshall Hall, who, like Aschbacher, was at the California Institute of Technology, called him 'The Steamroller'. He was the key player, leading the attack, while Gorenstein organized some of the other players into a team, communicating with group theorists on the East Coast, on the West Coast, in the Midwest, in Germany, in Britain, in France, and wherever there was anyone at the cutting edge in this type of mathematics. Gorenstein even tried to get the Russians involved when two Soviet mathematicians got permission to attend a conference in California.

But this was not a project that anyone could easily join. The technicalities were formidable, and Gorenstein readily admitted that this put other mathematicians off:

> Finite simple group theory was establishing a well-deserved reputation for inaccessibility because of the inordinate length of the papers pouring out. The 255-page proof of the [Feit–Thompson] theorem, filling an entire issue of the *Pacific Journal*, had set the tone, but it was by far not the longest paper. Moreover, the techniques being developed, no matter how seemingly powerful for the problems at hand, appeared to have no applications outside finite group theory. Although there was admiration within the

mathematical community for the achievements, there was also a growing feeling that finite group theorists were off on the wrong track. No mathematical theorem could require the number of pages these fellows were taking! Surely they were missing some geometric interpretation of the simple groups that would lead to a substantially shorter proof.

The view from inside was quite different: all the moves we were making seemed to be forced. It was not perversity on our part, but the intrinsic nature of the problem that seemed to be controlling the direction of our efforts and shaping the techniques being developed.*

As the huge effort increased, more and more young mathematicians got drawn in, and big international conferences were arranged to bring everyone together. Before getting on to those, let us return to Fischer and his monsters.

After Fischer had ceased being dean he found time to return to research and reconsider a possible symmetry atom generated by mirror symmetries in which the angle between any two mirrors was 90°, 60°, or 45°. His efforts were not in vain and in the summer of 1973 he managed to snag a huge one, the largest so far, of size 4,154,781,481,226,426,191,177,580,544,000,000. At this point in the book we may have become accustomed to large numbers because these symmetry groups can contain a vast number of operations, while the system of objects (mirrors, vertices, or whatever) is of a far more modest size. But this new group needed 13,571,955,000 mirrors, which makes Fischer's other monsters seem small by comparison.

There are two questions here. One is how on earth Fischer worked out these numbers, and the other is how such a vast system could ever be constructed. Let's first deal with how the numbers were worked out. Here is the calculation for the number of mirrors:

$$1 + 3,968,055 + 23,113,728 + 2,370,830,336 + 11,174,042,880 = 13,571,955,000$$

Let's look briefly at where these numbers come from. Fix one mirror – that is the number 1 at the beginning of the addition sum. Each of the other mirrors has an angle of 90°, 60°, or 45° with this one mirror. Those at 90° split into two sets whose sizes are the next two numbers in the sum. The fourth number is the number of mirrors at 60° to the one you fixed, and the fifth is the number at 45°. Fischer calculated each of these numbers as the size of one symmetry group divided by the size of another, and that yielded each number as a product of prime numbers, which had to be multiplied together before doing the addition. This was back in the early 1970s when there were no pocket calculators. People who had to do lots of additions and multiplications – people in accountancy firms, for example – would call in a comptometer operator, someone who sat down at a rather unwieldy machine and banged in the numbers at a great rate of knots.

Fischer was doing the work that led to these numbers in 1973 and was frequently in England, visiting the Universities of Warwick and Birmingham. In the autumn he visited Cambridge, where Conway tried to use an old mechanical calculator to help do the final calculations; but he couldn't find the parts, so Fischer and his wife did the calculations by hand. Doing it by hand there was more chance of a mistake, but as Fischer says, 'I knew the answer had to be divisible by 31, so there was a built-in check'. Once Fischer had the total number of mirrors he could multiply it by a number he already knew, which was the size of the sub-group fixing one mirror, and that would give him the size of the group – the whole symmetry atom.

In September there was a mathematics conference on finite groups at Oberwolfach, that wonderful retreat in south-western Germany, and Fischer spoke on his new group. This was an exciting event. By far the largest symmetry atom ever discovered had just been announced, but Graham Higman from Oxford was in Australia and wasn't there to hear it, so they sent him a card. One of his colleagues from Oxford said, 'If you want to make sure he reads it, keep it short. Just write down the size of this new group.' And they did just that. When mathematicians write numbers, particularly in a case like this, they factorize them into prime factors; for example, $24 = 2^3 \times 3$, and $60 = 2^2 \times 3 \times 5$. When they wrote to Higman they factorized the number for him, and in that form it was $2^{41} \times 3^{13} \times 5^6 \times 7^2 \times 11 \times 13 \times 17 \times 19 \times 23 \times 31 \times 47$.

The following month there was a second conference, this time at the University of Bielefeld, where Fischer was a professor. Thompson, Conway, Aschbacher, and others were there, and since Fischer's new group was the hot topic, everyone pressed him to give at least one talk. But there is a tradition in Germany that the host *never* gives a talk, so Fischer could only have informal discussions with the guests. This was quite a strict rule, and at a group theory conference in Germany a few years earlier, the organizers did not have enough willing speakers, so one of them gave a lecture himself. One senior participant immediately walked out.

Having worked out the size of the new group, and the number of mirrors, there was a very important question to answer. Did it really exist? If so it would permute 13,571,955,000 mirrors among themselves, and the question was how to construct such a vast system of permutations. I mentioned earlier that computer methods were used for some of the strange symmetry atoms that had been 'discovered' by the cross-section method, and it was natural to try using them here. These computer methods constructed the symmetry atom as a group of permutations, but you needed plenty of technical information, so Fischer set about working out the character table.

At the same time, he noticed that this massive group of permutations might be a cross-section in something even larger. This was in late 1973, and Bob Griess at the University of Michigan had a similar idea. Fischer and Griess were both convinced that Fischer's huge new group could appear as a cross-section in a larger group. How large, no one quite knew – at this stage things were still a bit hazy.

Let's remind ourselves how we got here. A few years earlier, Fischer had created his 'transposition' groups $Fi22$, $Fi23$, and $Fi24$. He had called them $M(22)$, $M(23)$, and $M(24)$, because they were related to Mathieu's groups $M22$, $M23$, and $M24$, and since he used $Fi22$ to create his new group of mirror symmetries, he tentatively called it M^{22}. It seemed to appear as a cross-section in something even bigger, and as this larger group was clearly associated with $Fi24$, he labelled it M^{24}. Was there something in between that could be called M^{23}? Fischer visited Cambridge to talk on his new work, and Conway named these three potential groups the Baby Monster, the Middle Monster, and the Super Monster. When it became clear that the Middle Monster didn't exist, Conway settled on the names Baby Monster and Monster, and this became the standard terminology.

Fischer had worked out the size of the Baby Monster, and

he was very keen on working out its character table. The Monster at this stage seemed out of reach – even its size was unknown. This was calculated partly by Fischer and partly by people at Cambridge, so let us turn to the work that they were doing there.

Thompson and Conway were in Cambridge, and the last we heard they had pulled some exceptional symmetry atoms from the Leech Lattice. Conway had discovered three new ones, and the lattice had yielded a total of 12, nine of which had already appeared elsewhere. These 12, along with Janko's groups $J1$ and $J3$, which had nothing to do with the Leech Lattice, brought the total number of exceptions to 14. This was in 1968. By the end of 1972 there were six more: three found by Fischer; one by Dieter Held, who was a colleague of Janko's in Australia, and later moved back to Germany; one by Richard Lyons, a student of Thompson's in the USA; and one by a man in the USA named Arunas Rudvalis. Both Held and Lyons used the cross-section method. Rudvalis used permutations; he found some solid evidence, after which there was a race between Conway and David Wales at Caltech on the one hand, and Griess on the other, to construct the permutations that were needed. Conway and Wales won the race. The total number of exceptions was now 20. With so much activity and so much information coming in, it made excellent sense to collect it all, correct errors, and present it in a form that was easy to read and readily available.

A new project, initiated by Conway and named 'the Atlas', was born. It came about as follows. Conway had a student named Robert Curtis who had written a thesis on the sub-groups in Conway's group $Co1$. In 1972 Curtis returned to Cambridge from a year at the California Institute of Technology, and Conway recalls that 'I applied for a grant for three years to start work on the Atlas, with Curtis as assistant.' Curtis was delighted to accept this position, and his office became the Atlas office. 'We called it "Atlantis" ', said Conway, 'because everything disappeared without trace.' They also used the word 'Atlantic' because the Atlantic ocean was named after the Atlas mountains in North Africa, which in turn were named after Atlas the Titan from Greek mythology. In the Atlas office they used blue paper, Atlantic blue.

As the project developed, Conway and Curtis were joined by others. Simon Norton was fascinated by the work, and kept popping in to see how things were going. Conway was initially disconcerted by this frequent visitor, but they soon realized how important Norton could be, so within a few weeks of his regular appearances they invited him to join the team. Norton had come to Cambridge straight from one of the top boarding schools in England, where he had been so extraordinarily good at mathematics that they put him in for a University of London degree, which he received when he left his boarding school. He then went to Cambridge to do the equivalent of a master's degree, followed by a PhD. Conway recalls that Norton was amazing. 'He seemed to absorb anything you taught him at a fantastic rate.'

The materials for the Atlas were kept in a binder labelled ATLAS. It gradually expanded, and eventually burst, so they covered it with a fake leather covering from one of the chairs in the mathematics common room, binding it on with a shoemaker's bradawl. The ATLAS binder collected technical information on all the exceptional symmetry atoms, as well as some other symmetry atoms in the periodic table. In 1973 two new exceptions were discovered, one being Fischer's Baby Monster and the other a new group of symmetries found by Michael O'Nan at Rutgers

University – the place where Gorenstein worked. This brought the total to 22, though not all these groups were yet known to exist. If a group emerged from the cross-section method – and this is how the Baby Monster appeared – then a great deal of information needed to be calculated before a construction was possible. Most of this data was encoded in the form of a square array of numbers called a character table. Character tables cropped up earlier when we were dealing with some of Janko's symmetry groups, and it is now time to explain what they are.

A character table is a square array of numbers. For example, the group of all permutations of four beads has a character table with five rows and five columns. This group has 24 operations in total, but they come in five different types,* and there is one column for each. The rows of the character table express the fundamentally different ways the group can operate in multidimensional space – any multidimensional operation can be obtained by combining these. The number of rows always equals the number of columns. When a group is composite – built up from many small cyclic groups, for example – its character table can have thousands of rows and columns. But symmetry atoms are different. For example, Mathieu's largest group has size 244,823,040 but its character table has just 26 rows and 26 columns. Each type of operation, except the one that does nothing, appears many times, so the symmetry atom can be extremely big while the character table is relatively small. We have not yet seen the size of the Monster, but although it has more operations than there are atoms in your personal computer, its character table has only 194 rows and columns.

The Atlas workers were accumulating character tables, along with other interesting facts about these symmetry atoms, but they didn't just *collect* the information, they checked it in great detail, and as Curtis said later, 'So many of the character tables we inherited had mistakes in them.' After correcting errors, and working out new details, they finally went for publication in 1985, and were proud to claim that their character tables were the cleanest in the world.

There are several ways to check character tables. For example, there is a way of combining two rows to give a single number; calculating this number can take a while if the character table is large, but in the end it has to be zero – if it isn't there is an error. Calculating with the many entries in these character tables needed some serious computing ability and they took on a new member named Richard Parker, who was terrifically good at this.

Now they were four, but a fifth member joined, named Robert Wilson. He made a speciality of working out the sub-groups of each exceptional symmetry atom. Some sub-groups are contained in others, so Wilson concentrated on those that were not contained in anything larger – they are called maximal, and Conway dubbed him Mr Maximal Subgroups. Now there is a very odd thing about the names of the Atlas authors. Here they are:

```
J. H.   C  O  N  W  A  Y
R. T.   C  U  R  T  I  S
S. P.   N  O  R  T  O  N
R. A.   P  A  R  K  E  R
R. A.   W  I  L  S  O  N
```

Notice that each surname has exactly six letters, and its vowels are always in second and fifth positions. The order of the names is the order in which they each joined the Atlas project, and remarkably enough this is the same as the alphabetical order. Rob Curtis, who is now at the University of Birmingham, also pointed out to me that in the Birmingham telephone directory these names are in decreasing order of rarity, Conway being the rarest and Wilson the most common. The Atlas members loved this sort of word play, and I am convinced that if someone with a name like Wolsey had come along to do a PhD in mathematics at Cambridge, he would have been welcomed to join the project as a sixth member, provided he had two initials of course.

The Atlas project was one of the most unusual ever in mathematics. It had a long gestation period, because of the detailed calculations involved and the fact that new symmetry atoms were still being discovered. It was finally published in 1985, by Oxford University Press.

In late 1973, while the Atlas work was in its early stages, the Monster was just peeping above the horizon, and the first thing to be calculated was how big it was. Fischer was working on this and visited Cambridge. He knew that the Monster had two cross-sections, and using these, along with a procedure invented by Thompson – called the Thompson order formula – the size of the whole thing was within reach. Thompson's technique needed detailed computations on how the two cross-sections could intersect one another, but even without having perfect information, Fischer used it to show that the size could not be greater than a certain number. Further calculations showed that the size had to lie in various arithmetic progressions, and after Fischer had gone back to Bielefeld, Conway used a programmable HP65 calculator to see what the smallest possibility was. He left the calculator running all night, and in the morning it gave him a number. He guessed this was correct, so he immediately wrote Fischer a letter saying, 'Dear Bernd, the size of the Monster is . . . I suppose you know this by now.'

Fischer didn't know, but combined with what he did know, this number was no longer a guess. The size of the Monster was now known, and in the first week of January 1974, Fischer talked on the new group at a working conference organised by Baer in Oberwolfach. In a leather-bound tome called the *Vortragsbuch* (*Book of Talks*) he wrote a report, the first publicly available mention of the Monster:

> There appear to be simple groups [symmetry atoms] of the following sizes:
>
> G_1: $2^{41} \cdot 3^{13} \cdot 5^6 \cdot 7^2 \cdot 11 \cdot 13 \cdot 17 \cdot 19 \cdot 23 \cdot 31 \cdot 47$
>
> G_2: $2^{15} \cdot 3^{10} \cdot 5^3 \cdot 7^2 \cdot 13 \cdot 19 \cdot 31$
>
> G_3: $2^{14} \cdot 3^6 \cdot 5^6 \cdot 7 \cdot 11 \cdot 19$
>
> G_4: $2^{46} \cdot 3^{20} \cdot 5^9 \cdot 7^6 \cdot 11^2 \cdot 13^3 \cdot 17 \cdot 19 \cdot 23 \cdot 29 \cdot 31 \cdot 41 \cdot 47 \cdot 59 \cdot 71$
>
> . . . The sizes of G_2, G_3, G_4 were determined by Conway, Harada and Thompson.

In this quotation G_1 denotes Fischer's Baby Monster, and G_4 the Monster; G_2 was later named after Thompson, and G_3 after Harada and Norton, in honour of those who did most of the calculations on them.

Having the size of the Monster was essential before working out its character table. This was calculated in Birmingham, not in Cambridge, but the Cambridge people made a vital contribution. The first row of any character table is trivial and just consists of a series of ones, but Simon Norton and others at Cambridge managed to figure out that the second row probably started with the number 196,883, which was the product of the three largest prime divisors in the size of the Monster. Certainly the number could not be less than this, and it meant that the smallest dimension the Monster could operate in was 196,883. This is pretty big, even for mathematicians, and it led to all sorts of amazing things, but first let us return to Fischer's work on the Baby Monster.

While the Atlas work proceeded in Cambridge, Fischer went to visit another mathematician with a similar outlook. Fischer betook himself to Birmingham in England to consult with Dr Livingstone. Donald Livingstone occupied a chair in Birmingham, having moved there five years earlier from the University of Michigan in Ann Arbor.

Livingstone had a rather interesting pedigree in mathematics. He did not go to school until he was 11, owing to his parents' severe financial problems. The family lived in South Africa, his father having moved there from the Isle of Mull on the west coast of Scotland after the First World War to try his luck at farming, but poor crops meant no money for schooling. Livingstone retained a great liking for Africa, and loved the Zulu language, which he spoke fluently.

Before settling in Birmingham he spent nine years at the University of Michigan in Ann Arbor. At that time he worked at home a lot, particularly in the late evening, and his youngest son remembers that 'Mathematics seemed to be something you did at night fuelled by coffee and cigarettes. On summer evenings in Ann Arbor he would sit out on the porch with paper, pencil, and cigarettes, and you could always win brownie points by taking him a cup of strong coffee.'

When Livingstone moved from Michigan to Birmingham he took some students with him, and they formed a small team working, as it were, on a second Atlas project, studying the exceptional symmetry atoms, finding sub-groups, and working out character tables. Fischer made lengthy visits to Birmingham, and he and Livingstone got on well because they had similar tendencies in mathematics. They both loved intense work on technical details, with an adequate supply of cigarettes and coffee, and both were rather diffident about writing up their results.

In 1974, when Fischer went to consult Livingstone about the character table of the Baby Monster, Livingstone had a different idea. The people at Cambridge had calculated that it could probably operate in 196,883 dimensions, so 'Livingstone said that using this we ought to be able to work out the character table of the Monster – then we could do the Baby Monster later.' So they started on this immense project.

Working out the character table of the Monster was hard. It required lots of calculations, so they needed computers, and eventually joined forces with a man named Mike Thorne who did the programming. 'He was really good at writing the programs, and was always available when we needed him.' This was back in 1974 when computers were nowhere near as powerful as they are today, and they needed to use the big machines at the University of Birmingham. Unfortunately some of the science departments needed big computing power every day, so you couldn't just log on to the computer whenever you felt like it. They had to wait until night-time, but Livingstone was a night owl and Fischer happily accommodated himself to the strange hours. They worked during the day too, and Fischer recalls that 'I was frequently at Birmingham in 1974. I would have six or eight weeks without administration, and I went there and worked 16 hours a day. We used the computer all the time, except Friday evenings.'

Fischer brought an important technique to this work. The Monster contains a cross-section involving Conway's largest group extended by a factor of over 32 million. It will appear later in constructing the Monster, and Fischer had a method of finding the character table of such a group. This one had over a thousand rows and columns, so he calculated it on a computer, and then they had to transfer it to the computer they were using for the Monster's character table. This would be a simple matter today. One would send it electronically across the university's network. In 1974 this wasn't possible – there was no university network – and it had to be done by hand. The same thing could happen today if the network were out of action – you would copy it to a disk, or to a solid-state storage device, and carry it over, but they had to copy it on to tapes. These were slow, and the whole operation took five hours.

Working 16-hour days, at least whenever Fischer was in Birmingham, it took them more than a year to complete the Monster's table. As Fischer recalls it, 'At the first go we had 18 characters [rows of the table]. Then it went up to 44 without my presence. Then there was a blockage.' The Birmingham people needed Fischer's presence, so 'I went to Birmingham for another short visit and gave a hint on how to get four more characters. After that a standard computation gave still more. After 70 or 80 characters the remainder were done by Livingstone using a completely different method.' The total number of characters was 194, and I remember Livingstone giving a talk on this character table at a conference. He brought in a ream of computer printout, with each of the 194 rows labelled by a Chinese character, as befits a mathematician who reads Chinese but runs out of alphabetic symbols – it is more fun than simply labelling them x_1, x_2, x_3 and so on, up to x_{194}.

Once these three musketeers, Fischer, Livingstone, and Thorne, had won the battle, and nailed the Monster's character table, Fischer went back to the Baby Monster. This was hard – in some ways harder than the Monster – but having the Monster's table at least made it reasonable. While Fischer was still working on the Baby Monster's character table, he was visiting various other universities, and in 1976 he spent three weeks at Rutgers. Charles Sims, he of the Higman–Sims group, worked there, and Fischer recalls, 'Sims told me his idea of how to construct the Baby Monster.' The idea was to create it on a computer as a group of permutations on 13,571,955,000 mirrors. The permutations were so enormous that Sims had to find a way of cutting them down so that the computer could handle it. 'Sims told me he could do it provided he had some further technical information, so I worked out what he needed. Then I left.' That summer, Jeffrey Leon from the University of Illinois at Chicago went to Rutgers for a year, and started collaborating with Sims on using a computer to construct the Baby Monster as a group of permutations. They succeeded and submitted their results for publication in February 1977.

Now it was the Monster's turn, but this was an order of magnitude more difficult. With the Baby Monster, Fischer recalls that 'They needed about 80 to 100 sub-groups, but with the Monster there would be over a 1,000. We decided not to try it if someone had a better idea.' And the Monster needed to permute far more mirrors than the Baby: 97,239,461,142,009,186,000 to be precise.

Computer methods seemed inadequate to the task, but the hand of man was sufficient. We will find out more about how it was done later.

In the meantime let's recall where we are. The Monster and two of its sub-groups, which were also new symmetry atoms, brought the total number of exceptions up to 25. But in 1975, Janko found the evidence for another one, later constructed by Norton and others at Cambridge. This brought the total number of exceptions to 26, where it has remained ever since.

The main excitement now was in proving there were no further exceptional symmetry atoms not in the periodic table. This was a time for large conferences to bring everyone together, and in summer 1978 a huge one took place at the University of Durham in England. By the end of summer 1978 most experts felt the Monster would prove to be the largest exception – that there were no more to be found. They were right – but there was a very big surprise to come.

15

A Monstrous Mystery

> Mathematical discoveries, like springtime violets in the woods, have their season which no human can hasten or retard.
>
> Carl Friedrich Gauss (1777–1855)

In scientific investigations that are nearing completion, one unexplained fact can suddenly open up a whole new area for investigation, and the classification and discovery of all finite symmetry atoms was a case in point. The source of exceptions had dried up, and the experts felt that proving the list was complete was just a matter of time. But the largest exception – the Monster – had unforeseen consequences. What happened was this. A British mathematician living in Montreal, named John McKay, was sitting at home one November day in 1978 reading a research paper. We met McKay earlier – he was a catalyst for Conway's work on the Leech Lattice when he and Leech were spending a year at the Atlas laboratory near Oxford. That was ten years earlier, and now he came up with something else that needed investigating. McKay is very eclectic, drawing inspiration from many sources, and the research paper he was reading was in number theory, the branch of mathematics that deals with the whole numbers. It was a paper by two British mathematicians, Oliver

Atkin at the University of Illinois at Chicago and Sir Peter Swinnerton-Dyer at Cambridge, and it discussed something called the j-function. McKay wanted to know more about this mysterious object, so he did a little reading and found there were several ways of specifying it. One of these gives the following series:

$$j(q) = q^{-1} + 196,884q + 21,493,760q^2 + 864,299,970q^3 \\ + 20,245,856,256q^4 + \ldots$$

McKay was astonished. The first significant number in this increasing sequence of coefficients is 196,884, and the smallest number of dimensions in which the Monster can act non-trivially is 196,883.

These numbers were too close for coincidence. McKay was excited and wrote a letter to John Thompson, the great guru of finite group theory. Rather than send the letter by post he gave it to Fischer, who was visiting Princeton and went to Montreal to give a talk. Thompson was also in Princeton at that time.

Another person in Thompson's position might have waved the coincidence away. After all, the j-function and the Monster came from different parts of mathematics, and any coincidence in the numbers might be meaningless. But Thompson was sufficiently open-minded to want to know more. Could the other coefficients of the j-function also be related to the Monster?

The first thing was to look at the Monster's character table, which has 194 rows and columns. A character table, as I mentioned earlier, is a square array of numbers that condenses a massive amount of information about the group in question. Each row expresses a fundamental way the group can act in a multidimensional space, and the first number in the row is the number of dimensions – it is called the character degree. The first two character degrees for the Monster are 1 and 196,883 – the first one represents a trivial action in one dimension, and the second represents a non-trivial action in 196,883 dimensions. Putting these together gives an action of the Monster in dimension 196,884. This number is the first significant coefficient of the *j*-function, and Thompson wondered whether the others might arise in a similar way. 'I started fooling around with it, and tried the next coefficient.'

The first few dimensions – or character degrees – of the Monster are shown in the column on the right below. Compare these with the coefficients occurring in the *j*-function on the left.

Coefficients for the j-function	Character degrees for the Monster
1	1
196,884	196,883
21,493,760	21,296,876
864,299,970	842,609,326
20,245,856,256	18,538,750,076

Simple addition shows the astonishing fact that by adding character degrees for the Monster you can get the first few coefficients for the *j*-function:

$$196,884 = 1 + 196,883$$

$$21,493,760 = 1 + 196,883 + 21,296,876$$

$$864,299,970 = 1 + 1 + 196,883 + 196,883$$
$$+ 21,296,876 + 842,609,326$$

This is more than a chance coincidence of two numbers. Thompson checked out more of them, and Fischer did likewise. The results were striking and word soon got round, though some people outside the field thought it was crazy – one even said he thought Thompson had finally lost it. On the contrary, Thompson was on to something, and this was not the first strange phenomenon associated with the Monster.

A few years earlier, another mathematician, named Andrew Ogg, from the University of California at Berkeley, had made an entirely different observation. Ogg had finished off an old problem related to the *j*-function, dating from the nineteenth century. It involved finding all the prime numbers that could be used to obtain other '*j*-functions', in a way I will explain later. These prime numbers turned out to be the following: 2, 3, 5, 7, 11, 13, 17, 19, 23, 29, 31, 41, 47, 59, 71. In January 1975, while spending an academic year in Paris, he attended the inaugural lecture for Jacques Tits, the man who invented multi-crystals (or 'buildings'). Tits had just moved from Bonn to Paris, to take up a chair at the Collège de France, and his inaugural lecture was a year after the Monster had been discovered. He mentioned it in his talk, writing its size on the board as a product of prime numbers:

$$2^{46} \times 3^{20} \times 5^9 \times 7^6 \times 11^2 \times 13^3 \times 17 \times 19 \times 23 \times 29 \times 31 \times 41 \times 47 \times 59 \times 71$$

Ogg was astonished. These were precisely the prime numbers playing a special role in the problem he had recently settled. He mentioned this extraordinary fact to Tits, and to Jean-Pierre Serre, a colleague of Tits who had written books on several areas of mathematics, including one called *A Course in Arithmetic*, involving the *j*-function in number theory. A young mathematician I know was once reading it on the New York subway, and a well-meaning lady came over to tell him how sensible he was as an adult to relearn the basics.

Serre was one of the top mathematicians of the century, but Ogg's observation was news to him, and his response was '*Sans blague*' (No kidding). What the reason was, no one had the faintest idea. Ogg wrote about this coincidence in a paper he was working on, and offered a bottle of Jack Daniels for an answer. When McKay made his own observation nearly four years later, the offer was still open.

In order to grasp the way Ogg's prime numbers arise we need a new concept. This takes us back to the days of the ancient Greeks.

In 300 BCE, Euclid of Alexandria wrote *The Elements*, a sequence of books laying out mathematics, as it was known at the time. It was a superb piece of work, and over a thousand years later was translated from Greek into Arabic, and three hundred years after that from Arabic into Latin. During the European Renaissance, Greek copies were found, and translated directly into Latin, and later into the individual languages of Europe. Learning geometry at school was often referred to as learning Euclid, and his exposition was wonderful. He started by stating a set of axioms and went on to prove theorems that are as valid today as they were in 300 BCE.

Euclid's axioms for geometry in a plane are usually expressed as five statements, but the only one I want to discuss here is the fifth axiom – the one about parallel lines. This has the effect of saying that if you take a straight line L in the plane, and a point p not on that line, then there is exactly one line through p that does not meet L however far these two lines are extended in either direction. Two such lines are usually called parallel.

Later mathematicians, both in the Middle East and Europe, tried to prove that Euclid's fifth axiom was a consequence of the other four axioms. Some of these 'proofs' were quite sophisticated, but all of them were wrong. Eventually it was shown that Euclid's fifth axiom could not be proved, because there is a 'non-Euclidean' plane in which it fails. This was discovered independently by a Hungarian mathematician, Janos Bolyai, in the 1820s, and a Russian, Nicolai Lobachevski. It satisfies the other four axioms, but is distinguished from the Euclidean plane by the fact that the angles of a triangle add up to less than 180°: The larger the triangle, the smaller the sum of the angles – as its vertices head out to infinity the sum of its angles approaches zero.

Janos Bolyai's father, Farkas Bolyai, had worked on the problem of parallel lines himself, and was alarmed that his son had decided to devote his attention to it. He wrote to him,

> You must not attempt this approach to parallels. I know this way to its very end. I have traversed this bottomless night, which extinguished all light and joy of my life. I entreat you, leave the science of parallels alone.*

However, Janos persisted, and in 1823 was able to tell his father that he was succeeding. 'Out of nothing I have created a strange new universe.' By 1831 he had written it up as a 24-page appendix to a two-volume treatise on mathematics written by his father, who, being a friend of the great Karl Friedrich Gauss, proudly sent it to him, expecting praise for his son's great achievement. What he got in response was:

If I commenced by saying that I am unable to praise this work, you would certainly be surprised for a moment. But I cannot say otherwise. To praise it, would be to praise myself. Indeed the whole contents of the work, the path taken by your son, the results to which he is led, coincide almost entirely with my meditations, which have occupied my mind for the thirty or thirty-five years . . . my intention was not to let it be published during my lifetime.*

The young Janos Bolyai was disheartened by Gauss's response, but it hasn't affected our later admiration for his achievement, nor that of Lobachevski, which was done independently. Lobachevski was a very active mathematician and administrator who spent his whole career at the University of Kazan, first as a student and later as its rector. As for Gauss, his manuscripts and letters, published after his death, showed that he had indeed discovered non-Euclidean geometry independently.

The Bolyai–Lobachevski plane, usually called the hyperbolic plane, is not as easy to envisage as the Euclidean plane. Mathematicians regard it as a surface of negative curvature, as opposed to the positive curvature of a sphere, or the absence of curvature in the Euclidean plane, but negative curvature is hard to imagine. The positive curvature of a sphere is easier, because when you flatten it out, as we do when we draw a map of the world, things that are further from the centre appear larger than they really are, so at least we can see them clearly. For example, on most maps of the world Greenland looks much larger than any country in Africa, whereas in reality, Algeria, Congo, and Sudan are all larger than Greenland. In the hyperbolic plane the opposite happens. As you move outwards from the centre, things appear smaller than they really are.

There are several ways of picturing the hyperbolic plane, but the most elegant is the Poincaré model, named after a famous French mathematician of the late nineteenth century, Henri Poincaré. He viewed it as a disc in which distances get increasingly foreshortened as one approaches the boundary, and where straight lines only appear straight if they go through the centre of the disc. All others appear as arcs of circles meeting the boundary of the disc at right angles, as in the picture below. The Poincaré model has the advantage that although straightness of lines is not preserved, the angles between them are unchanged.

With the hyperbolic plane to hand we can return to Ogg's prime numbers. He was working in number theory with something called the modular group, which allows one pair of integers (an integer is a positive or negative whole number) to change into another. The modular group operates on the hyperbolic plane, rolling it up onto a sphere.

Ogg was looking at groups derived from the modular group – one for each prime number. These groups roll up the hyperbolic plane less tightly than the whole modular group, yielding larger surfaces that are not necessarily spheres. They are two-sided surfaces that can be deformed, without tearing or pasting, to resemble either a sphere, or a torus, or a double torus, etc. The mathematics dealing with such things is called topology, and these surfaces are classified by their topological genus: a surface like a sphere has genus 0, one like a torus has genus 1, like a double torus, genus 2, and so on.

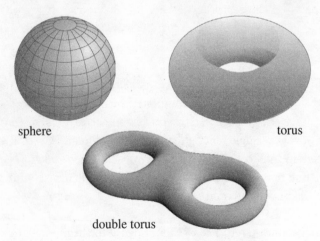

sphere   torus

double torus

Ogg had proved this surface was a sphere precisely when the prime number is one of: 2, 3, 5, 7, 11, 13, 17, 19, 23, 29, 31, 41, 47,

59, 71. These were exactly the primes dividing the size of the Monster, a fact having no explanation at all, and perhaps simply an odd coincidence.

There now seemed to be two strange connections between the Monster and number theory. One was the j-function, which according to Thompson's calculations seemed to be related to the Monster's character degrees, and the other was Ogg's set of prime numbers. Of course the coincidence of the primes could be accidental; there were not many of them, they were relatively small, and if you added one to each of the three largest, 47, 59, and 71, you obtained multiples of 12, a number playing a special role in the Monster. Ogg's observation was interesting, but not worth pursuing on its own. Fortunately Thompson's numerology between the Monster and the j-function involved much larger numbers, which seemed less accidental. Moreover, it helped to substantiate the coincidences Ogg had noticed because there was an important connection between the modular group and the j-function, as any number theorist worth their salt could tell you.

When you wrap the hyperbolic plane on to a surface using a group derived from the modular group, you get an algebraic structure on the surface. When the surface is a sphere, this structure is generated by a single function, and for the modular group it is the j-function. In effect the modular group yields the j-function, and in a similar way the groups for Ogg's special prime numbers yield what I will call mini-j-functions*.

When Thompson returned to Cambridge at the beginning of 1979 he explained to John Conway that by adding character degrees in the first column of the Monster's character table he could get the first six coefficients of the j-function, and remarked, 'If you try other columns you might get some interesting series.' Conway had the Monster's character table, as worked out by Fischer, Livingstone, and Thorne – it was part of the great Atlas project – so he was keen to try this out. He started adding together numbers in the second column of the character table, just as Thompson had done with the first column. Then he tried other columns, obtaining various series whose coefficients were increasingly large numbers. One of the early numbers he pulled out was 11,202, which was easy to remember and seemingly unrelated to anything interesting. He went to the library to consult nineteenth-century papers on number theory, and when he found this number appearing in a series in one of those papers he was convinced he was finally on to something. Writing about it later he recalled that 'One of the most exciting moments of my life was when, after computing several of these series, I went down to our mathematical library and found some of them in Jacobi's "Fundamenta nova theoriæ" . . . with the same coefficients down to the last decimal digit!'*

Conway was good at reading these old papers, and recalls that, as an undergraduate, 'I turned the pages of every paper Euler published in the new journal from St Petersburg' – Euler, who lived in the eighteenth century, was the most prolific mathematician of all time, and his proofs were very stimulating, with lots of good ideas. As Conway says, 'Euler would prove a theorem, and later someone would modify his theorem and produce a more complicated proof, but if you wanted to understand what was really going on you had to go back to Euler.'

Simon Norton became very interested too, but Conway recalls that 'Simon was travelling around the country on trains, so I got a two weeks' start on him, which was great because he was always so fast at learning anything new.' Simon was a great train aficionado, and regularly carried detailed timetables with him. I remember once at Oberwolfach asking his advice on how to get from one place to another in Germany, and he immediately pulled out a huge book to check the information.

Conway and Norton worked fast: 'It was such a big job, and took so much calculation that we worked on it day and night for six solid weeks.' Verifying that the earlier observations were no mere coincidences was hard work, and as Conway says, 'Observations are easy, and the information content is interesting, but the work involved in making them is insignificant by comparison with what we did. We were the first people to show this wasn't an accident.'

After an intense six-week period, doing thousands of computations, they had come up with real evidence substantiating Thompson's observations.

Meanwhile Thompson wanted to *prove* that these new series Conway and Norton had found would agree with the mini-j-functions at all coefficients. There were infinitely many coefficients, and Thompson had the idea of using a theorem of Brauer, in a way I explain below, but at this point he needed some help because he wasn't a number theorist. He wrote to Serre in Paris, who had heard first-hand from Ogg about the strange coincidences with prime numbers. Serre wrote back advising him to write to Oliver Atkin at the University of Illinois at Chicago.

Atkin was an expert on the j-function and the mini-j-functions. He was also a computer expert who had at one time worked at the Atlas laboratory, the place where Leech and McKay were working when they tried to interest people in Leech's Lattice. Many years before that, when he was a young man, Atkin had worked at Bletchley Park, the British decoding centre during the Second World War. Before long letters were flying back and forth between Cambridge and Chicago, and two of Atkin's group theory colleagues, Paul Fong and Stephen Smith, soon got involved.

Paul Fong had been a student of Richard Brauer, he of the big cross-section theorem that inspired Thompson and Feit, and in March 1979 Thompson wrote him a letter that started as follows:

> Dear Paul,
>
> How I wish Richard were alive! He would hugely enjoy what is happening. Among other things, there is an opportunity to use his characterization of characters in the novel situation of knowing the entire character table.*

Unfortunately Richard Brauer had died within the past two years. His 'characterization of characters' was a result that helped in working out character tables, but Thompson's idea was to use it in a situation where the whole character table was already known. The point was that Conway and Norton were adding entries in each column, and treating all columns the same way. This amounted to adding whole rows to one another. For example, one coefficient in each series was obtained by adding the first three entries in a column, and doing them all together meant adding the first three rows. The rows are the characters themselves, hence the need for Brauer's theorem. Thompson used

Atkin's knowledge of the *j*-functions, along with Brauer's characterization of characters, to reduce the whole problem to a finite number of calculations that Atkin did by computer. This proved that all the series Conway and Norton had produced, using different columns of the character table, were combinations of Monster characters.

Thompson wrote a couple of fairly short papers on his recent work, and John Conway and Simon Norton wrote a more expansive paper with the title 'Monstrous Moonshine'. It dealt with all columns in the Monster's character table, and showed how some could be obtained from others by the 'replication formulas' they invented. It was a detailed and technical piece of work, demonstrating a clear connection between the Monster and number theory.

The term Moonshine, like the name Monster, was suggested by Conway, and has a variety of meanings. It can refer to foolish or naive ideas, but also to the illicit distillation of spirits (particularly corn whiskey, from the days of prohibition in America). It gave an impression of dabbling in mysterious matters that might be better left alone, but also had the useful connotation of something shining by reflected light. The true source of light is probably yet to be found, but there were further strange connections to come later.

In the meantime other group theorists were hearing about all this at second hand, and it was time for another big conference. The previous one was in summer 1978 at Durham, and in summer 1979 the group theorists met again, this time at Santa Cruz, a campus of the University of California. This conference was unusual in having number theorists and group theorists together.

The point was to discuss the strange connection between the Monster and the *j*-functions, but the underlying reason for this connection remained – and still remains – elusive. We shall come back to this later. In the meantime, at the end of the 1970s, the existence of the Monster was still an open question. No one had yet constructed it, so let us turn to the problems involved.

16

Construction

Everything should be made as simple as possible, but not simpler.

Albert Einstein

In early 1977, when Sims and Leon had constructed the Baby Monster on a computer, as a group of permutations, it was natural to ask whether the Monster could be constructed in a similar way. Unfortunately this seemed out of sight, as I mentioned earlier, so an alternative method was needed. Perhaps one could use multidimensional space. Similar methods had been applied to other exceptional groups, such as Janko's first group $J1$. But where $J1$ needed seven dimensions, the Monster needed nearly 200,000, which means that a single operation in the Monster would appear as a matrix with nearly 200,000 rows and as many columns. As with $J1$ it might suffice to use two such matrices, and multiply them together in many different combinations, but when Fischer was first working on the Monster he reckoned that the time required for just one matrix multiplication would be about half a year of computer time.* Parallel processing with modern machines would be far quicker, but would still take many days, and that is just to do a single multiplication of two matrices. A mathematician could certainly be forgiven for thinking that a construction of the Monster wasn't worth wasting time on.

By the end of the 1970s the Monster was still not known to exist, despite all the technical information they had calculated. Looking back on those days, Conway said recently, 'At the time, I regarded its construction as a practical impossibility, or maybe an impractical possibility.' Then, rather suddenly, on 14 January 1980, Bob Griess at the University of Michigan at Ann Arbor announced a construction. Conway recalls that 'We had no idea how he'd constructed it when we got his card. We thought he must have done it using some new method, because I regarded the obvious way as too heroic.'

Bob Griess was one of the first people to come up with evidence for the Monster, back in November 1973. He had heard of Fischer's Baby Monster and calculated that there must be something beyond it. Fischer knew that too, and in collaboration with Livingstone had obtained the entire character table of the Monster, assuming the existence of an action in 196,883 dimensions. Simon Norton at Cambridge then used the character table to figure out that the Monster must preserve an algebra structure in 196,884 dimensions. This structure would allow any two points to be multiplied together to give a third point.

Griess's first task was to construct a suitable multiplication. This means he had to give rules that would assign to any two points, p and q, a third point r that would be their product, but one serious problem was that although he would often know that this product must be either r or $-r$, he wasn't sure which one.

Griess had occasionally thought about this problem, and in summer 1979 he decided to have another look at it. 'I kept massaging it, and I found I could begin to understand the sign problems by tracking them back into the group.' The group Griess is referring to here is a huge sub-group of the Monster – Conway's largest 'simple' group, extended by a factor of over 32 million. This group needs 96,308 dimensions, and is one of the two cross-sections of the Monster (the other one involves the Baby Monster). Fischer used it earlier in helping to build the character table of the Monster, and Griess now used it to help to construct the Monster in 196,884 dimensions. He knew that the action of this huge subgroup must split the space into three sub-spaces of the following dimensions:

$$98,304 + 300 + 98,280 = 196,884$$

The first number $98,304 = 2^{12} \times 24$. This is the space needed for the cross-section mentioned above.

The second number $300 = 24 + 23 + 22 + \ldots + 3 + 2 + 1$. This comes from a triangular arrangement of numbers with 24 in the first row, 23 in the second row, 22 in the third row, and so on – giving a total of 300. These numbers can be varied independently of one another, so there are 300 variables, or geometrically speaking 300 dimensions.

The third number $98,280 = 196,560 \div 2$. This comes from the Leech Lattice, where there are 196,560 points closest to a given point, and they come in 98,280 diametrically opposite pairs. Each pair yields an axis through the given point, and in the 196,884-dimensional space these axes become independent of one another.

In the summer of 1979, Griess became interested in trying to sort out the sign problem for the multiplication. But even if he could solve it, he still had to show that the group of symmetries contained the Monster, and the only visible group of symmetries he had was one of its cross-sections. Conway had a similar problem with the Leech Lattice, where the vertices nearest a given vertex were split into three sets. He had a large group of symmetries permuting the points of each set among themselves, but needed one extra permutation to create a larger group. Bob Griess now had a similar problem: his space was split into three sub-spaces and he needed one extra symmetry operation that would mingle these sub-spaces. If he could get it, he would almost certainly have created the Monster itself, though he would still have to prove that.

He realized that success would require a serious time commitment. You don't solve a problem like this by spending a couple of hours a day working on it, and there were other attractive things to work on, possibly offering a more likely chance of success. As Griess himself says, 'In the summer of 1979 the Moonshine stuff interested me. I had a quiet time playing with it, but then I considered the sign problem again. Fall came, I went to the Institute at Princeton for a term, decided to work more on the Monster construction, and just got addicted to it.'

Griess had married in June 1979, shortly before the big conference in Santa Cruz, and his new circumstances evidently inspired him. 'I was newly married, but in the spirit of being newly married she was very understanding. In October I just started working round the clock. I took off half a day for Thanksgiving and one day for Christmas.'

Griess was working on two interrelated problems: the sign problem for the algebra structure, and the creation of an extra symmetry operation. 'I started by trying to solve both . . . at the same time.' The sign problem for the algebra he solved first, but finding an extra symmetry also involved sign problems, and he recalls that 'it was a slippery business'. This was an exciting and exhausting time, and by mid-December he was very close, 'but each complete check was so long and mentally tiresome that I was not confident until a bit after the New Year'. By the middle of January, 'after a final unhurried check', he was ready to send out an informal announcement, on 14 January 1980. Writing up all the details was to take a lot longer, and a paper containing these details was only submitted for publication in June 1981. Such an important result went through the publication system as quickly as possible, but it needed careful checking by an external referee, and was a very long paper. It appeared in print in 1982, as 102 printed pages of detailed argument. In this paper, Griess uses the term Friendly Giant, rather than Monster, but his new name never caught on.

Since the Monster had such fascinating connections with other things, it wasn't long before two other mathematicians got deeply involved in looking at the Griess construction. One was Jacques Tits, who found a way of avoiding the sign problems. The judicious guesses that Griess made, which then had to fit together nicely, could be eliminated in favour of a guaranteed method. Tits also found a number of other improvements: 'I simplified Griess's construction somewhat, but what he did was a great piece of work. It was done without computers and this was a great feat.'

Indeed, the fact that Griess constructed the Monster without using a computer in any way was magnificent. After all, the

Monster was discovered using the cross-section method, and in all but one other case of this type, an eventual construction used computers. The one exception was Janko's group $J2$, which Marshall Hall and Jacques Tits had both constructed by hand as a group of permutations on 100 symbols.

The other person who took a detailed interest in the Griess construction, and then gave a construction of his own, was Conway. He calls Griess's construction 'monumental', comments on Tits's simplifications, and compares his own approach to that of Tits.

> Tits avoids the need for explicit consideration of the sign problems by a more abstract discussion of the underlying representations. He also has a very elegant proof of finiteness. In a sense, Tits's improvements are orthogonal to ours. He wishes to avoid all calculations in the Monster, while we would like to make it easy for the reader to perform such calculations for himself.*

Conway's construction is similar to Griess's in the sense that they both use the same large cross-section of the Monster to get started. This splits the 196,884-dimensional space into three sub-spaces, as I mentioned earlier. Griess had constructed a multiplication among the points of this huge space, and then obtained a symmetry that wasn't in the cross-section, allowing points from the three sub-spaces to interchange with one another. Conway avoided both these problems by creating three identical spaces of 196,884 dimensions, and merging them into one in such a way that corresponding sub-spaces took up different parts of the whole space. This trick meant that he constructed three identical looking cross-sections of the Monster, which between them generated the whole thing.

Conway's paper was published in 1985, the same year as the Atlas, and at this point it seemed to everyone concerned that all the finite building blocks for symmetry – the exceptional symmetry atoms – had now been found. There was, however, always the question of errors, and Conway reports that someone asked him about this, and whether he was an optimist or a pessimist.

> I replied that I was a pessimist, but still hopeful, and was delighted to find that this answer was misinterpreted in exactly the way I had maliciously desired!
>
> Among those who are engaged in the great cooperative attempt to classify all the finite simple groups, 'optimism' usually describes the belief that there are no more such groups to be found, since new groups appear as obstacles in the path of progress. My own view is that simple groups are beautiful things, and I'd like to see more of them, but am reluctantly coming around to the view that there are likely to be no more to be seen.*

By the time Conway wrote this, the question of *proving* that there are no more exceptions had already taken a new turn. Danny Gorenstein, in collaboration with his colleague Richard Lyons at Rutgers, and Ron Solomon at Ohio State University, started a project called the Revision. The idea was to put in place a complete proof of the Classification, readily accessible to posterity. This would make it possible for a new generation of mathematicians, not weaned on the enormous technicalities, to understand it all. It was a tough call. Many of the earlier papers were horrendously difficult to follow, and the Revision project was a very brave one indeed. It is still in progress.

When they started this project, Michael Aschbacher and others were nailing down most of the loose boards, after checking that nothing was hidden underneath, but some mathematicians outside the charmed circle still had a feeling of unease. It seemed that the group theorists had been racing along too fast, and could well be missing things. Some were annoyed by the apparent hubris, and I remember saying to one outsider that maybe they would yet find another exception, to which he fervently responded, 'I hope they find a whole family of them! I pray for it.'

Certainly there was cause for doubt. Geoff Mason at Santa Cruz in California had been working on the 'quasi-thin problem' – an essential part of the Classification programme – and seemed to have shown there was nothing new in this direction. Those who had seen drafts of his manuscript confirmed that the entire thing was very long indeed – about 800 pages of typescript – but it was not yet in a form for publication, and turned out later to have gaps in the argument. Writing in 1995, Solomon says:

> The literature on the Classification was always challenging, coming in massive 200-page papers. Nevertheless, there were always individuals and seminar groups that made serious efforts to read and digest most of the papers which appeared during the years 1960–1975. At least 3000 pages of mathematically dense preprints appeared in the years 1976–1980 and simply overwhelmed the digestive system of the group theory community. Mason's 800-page quasi-thin typescript has achieved some notoriety, inasmuch as it has never been published.*

Mason was trying to do the same as most other people, namely to close things off, rather than find anything new, and if a particular situation led to a contradiction, then that settled it and they could move on to other things. However, some contradictions were chimeras. They didn't really exist, and as Conway wrote in 1980:

> Quite a large number of the groups ... [were] constructed after somebody had already proved them impossible! When David Wales and I set out to construct the Rudvalis group, for example, we soon ran into a contradiction which refused to go away even after we had condensed it onto one side of a sheet of paper and scrutinised it for several days. Fortunately we were so convinced that the group existed that eventually we just put that piece of paper aside and constructed the group by another method that carefully went nowhere near our contradiction! Another group theorist later told me that he too had disproved the Rudvalis group, although he had only used the assumption that it contains a subgroup that it does, in fact, contain! ... What worries me is the nagging thought that another group like the Rudvalis group might have been disproved somewhere in the classification programme by someone who had no overwhelming conviction that it existed.
>
> The trouble is that groups behave in astonishingly subtle ways that make them psychologically rather difficult to grasp. We might say that they are adept at doing large numbers of things well before breakfast.*

When Danny Gorenstein died in 1992, Lyons and Solomon continued the joint Revision project, and expect to have the whole thing finished by 2010. As to the quasi-thin case, Gorenstein at first hoped that some new work, being done partly in Germany and partly in America, would settle it, but that turned out not to be the case, and the problem was still open.

Then at the annual American Mathematical Society meeting in San Francisco in January 1995, Michael Aschbacher from Caltech and Stephen Smith from the University of Illinois in Chicago organized a special session on the Classification. An unspoken goal was to find some enthusiastic young people to volunteer for quasi-thin, but that didn't happen.

They met again in May and, as Smith remembers, 'Michael suggested we should just bite the bullet and step up to the quasi-thin problem ourselves. We both did some preliminary thinking, and then started in earnest during January 1996 when I went to Caltech on a sabbatical.' Together they planned a book dealing with the quasi-thin stuff once and for all. Their book, *The Classification of Quasi-Thin Groups*, occupying over a thousand pages split into two volumes, came out in November 2004, and finally settled it.

Still some people wondered whether Janko, who had produced four exceptions, might not have a fifth one up his sleeve. Janko himself got in touch with Thompson to tell him where he thought another large exception might be lying, in quasi-thin territory, so Thompson called Smith in Chicago to make enquiries. However, it seemed they had that case well covered, and when I wrote to Janko to enquire what he now thought, his reply was, 'I have read ALL CRITICAL PLACES in the Aschbacher–Smith book (on quasi-thin simple groups), and now I am confident that the classification is absolutely OK!!' If Janko is confident, and if Thompson, Aschbacher, and others are confident, then that should give us all confidence. The proof of the Classification has come a long way from the time when a handful of experts believed in it, to the point where it is being written for future generations of mathematicians to understand. This is the role of the great Revision project, which will form a basis on which we can continue to strive for a better understanding of it all.

There remains, however, the great mystery about the Monster and Moonshine, about which I have more to say in the next chapter.

17

Moonshine

Thus the task is, not so much to see what no one has yet seen; but to think what no one has yet thought about that which everybody sees.

Erwin Schrödinger (1887–1961), a leading
discoverer of quantum theory

The Monster's connections with number theory – the Moonshine connections – had suggested it was a more beautiful and important group of symmetries than first realized, so there should be a more elegant way of obtaining it. The Griess construction in 196,884 dimensions, marvellous though it was, might then emerge within a broader picture, but before we look at this further, let us recall how the Monster was originally discovered.

The first big step was Mathieu's group of permutations $M24$, discovered in the mid-nineteenth century. A hundred years later this led to the Leech Lattice in 24 dimensions, which in turn led to Conway's group $Co1$, and then finally to the Monster. The sequence of symmetry groups is: Mathieu's group $M24$, Conway's group $Co1$, and the Monster.

As groups of permutations, one quickly finds that $M24$ permutes 24 objects, but $Co1$ needs to permute at least 98,280 objects. This is a big jump from 24, though it emerges naturally in

24 dimensions, as a set of axes through a vertex in the Leech Lattice. Each axis has two diametrically opposite points, which are the centres of spheres touching a given sphere. This gives $2 \times 98,280 = 196,560$ spheres touching a central sphere, which is the maximum possible in 24 dimensions.

Permutations are fine for $M24$, but not for $Co1$. It is the Leech Lattice that takes us from $M24$ to $Co1$, and we now need a good way of going from $Co1$ to the Monster. The fact that the number of *dimensions* climbs so sharply is analogous to the fact that the number of *points* increases sharply as we move from $M24$ to $Co1$. Taking this analogy further, notice that the Leech Lattice really moves us from 24 points to an infinite number, because a lattice stretches out to infinity in all directions; so as we move from Conway's group $Co1$ to the Monster, it seems reasonable to move from 24 dimensions to an infinite number. This is where the Moonshine connection comes in.

Following McKay's observation about the Monster and the j-function, Conway, Norton, and Thompson had shown that each coefficient of the j-function should be the dimension of a space on which the Monster acts. Putting these spaces together would give an infinite dimensional space. Conway and Norton conjectured that such a space should yield the mini-j-functions, along with their replication formulas, and these became known as the Moonshine conjectures.

The first significant coefficient of the j-function was 196,884 – the same as the dimension of the space that Griess created in constructing the Monster. An infinite-dimensional space for the Monster should surely start with this piece, or something very like it, and a few years later such a space emerged from the work of three collaborators: Igor Frenkel, James Lepowsky, and Arne Meurman. It had sub-spaces of all the correct dimensions for the *j*-function, and a group of symmetries that yielded the Monster. This was in 1984, and four years later, in 1988, they wrote a book on their work called *Vertex Operator Algebras and the Monster*. In the preface they write:

> This work grew out of our attempt to unravel the mysteries of the Monster, the most exceptional finite symmetry group in mathematics. The Monster creates a world of its own and many of the mysteries reflect the unity and diversity of this mathematical world. We began struggling with the Monster even before it was known to exist, as it was starting to reveal its true beauty. We have been able to solve some of the problems and to shed light on others, and we have added a few new ones.*

The vertex operator algebras in the title of the book were something fairly new. They had appeared a couple of years earlier as 'vertex algebras', but most mathematicians had never heard of them. Moreover, vertex operators originated not in mathematics, but in physics. They come from string theory, and describe the interaction of strings, which are models for elementary particles. This suggested a connection between the Monster and deep ideas from physics, and in the preface to their book, Frenkel, Lepowsky, and Meurman write, 'our main theorem can be interpreted as a quantum-field-theoretic construction of the Monster and in fact as the statement that the Monster is the symmetry group of a special string theory'. Before we pursue this theme, let us recall how we got here.

The Monster – the largest of the exceptional symmetry atoms –

had been shown to have deep connections with number theory, which Conway dubbed Moonshine. The first of these connections was with the *j*-function, and Conway and Norton then used the various types of operations in the Monster to produce a collection of mini-*j*-functions. They conjectured that the *j*-function along with these mini-*j*-functions should emerge from an infinite dimensional space having the Monster as its symmetry group, and a few years later, Frenkel, Lepowsky, and Meurman created a suitable space. They called it the *Moonshine module*, and although it yielded the *j*-function, it was not clear that it would also yield all the mini-*j*-functions. In other words, it was not yet known to satisfy the Moonshine conjectures of Conway and Norton, and the man who resolved this problem was Richard Borcherds.

When the Moonshine module was first announced in 1984, Borcherds was a postgraduate student at Cambridge, working under Conway, and he was hearing plenty about the Monster. Conway was publishing his own construction, and putting the finishing touches to the great Atlas project. The Monster was very much in the air, and Borcherds wanted to provide a new approach to it. He was an unusually capable student. Conway recalls that on one occasion he himself, his second wife Larissa (who was also a mathematician), and Richard Parker were all working on a problem concerning the Leech Lattice. The problem had started with some observations by Parker that Conway explained to Borcherds, and then continued working on with his collaborators. Six weeks later when Borcherds found they were still at it, he showed surprise: 'Oh, are you still working on that? I solved it some time ago.'

Solving problems is impressive, but Borcherds also liked to put things in a broader theoretical context, and shortly after his

PhD he published a remarkable paper on vertex algebras and the Monster. This linked up with the work of Frenkel, Lepowsky, and Meurman, and two years later, in 1988, he published a paper on an intriguing class of Lie algebras, which led him towards a proof that the Moonshine module satisfied the Conway–Norton conjectures.

Lie algebras go back to the work of Sophus Lie (see Chapter 5). His groups of continuous transformations – now called Lie groups – were classified by Killing and Cartan. Their work used algebraic structures, called Lie algebras, that have a particularly elegant form. They are based on crystalline structures that yield tight sphere packings – just like the Leech Lattice, but simpler. These crystal structures – or rather their symmetry groups – embed in the Lie groups, and the symmetry group of the Leech Lattice embeds in the Monster in a similar way. Perhaps by analogy the Leech Lattice could be used to obtain a Lie algebra that would yield the Monster.

This idea led Borcherds to a new class of Lie algebras, which he published in 1988 – they have since been called Borcherds algebras, or Borcherds–Kac–Moody algebras.* In a paper two years later, he presented a special case that he later called the fake Monster Lie algebra. This used the Leech Lattice in an intriguing way that is related to the mathematics behind special relativity theory, so let us return to the physics of Chapter 6.

In the first half of the twentieth century two major advances in physics had occurred: relativity theory and quantum theory. Einstein published his first paper on relativity theory in 1905, and a young German–Lithuanian mathematician named Hermann

Minkowski then produced a geometry that provided the perfect background for Einstein's theory. In Minkowski's geometry time and space are linked in a four-dimensional space-time. Each point in this space-time represents a possible event, and has four coordinates, three of space and one of time. The 'time-distance' between two points – in other words between two events – involves the differences in their four coordinates x, y, z, and t, where t is the time coordinate. In our usual space of three dimensions the square of the distance would be given by the formula $x^2 + y^2 + z^2$, but in Minkowski's geometry, the square of the 'time-distance' is:

$$x^2 + y^2 + z^2 - t^2$$

In this formula I am choosing units so that the speed of light is 1 unit. The important thing is the minus sign. This means that the square of the 'time-distance' between two points may be positive, negative, or zero. When it is negative – for instance, when x, y, and z are zero – the two points can be connected by an object travelling at less than the speed of light; when it is positive they cannot.

To illustrate a case where the square of the 'time-distance' is positive, imagine someone sending an e-mail message from a planet 100 light-years away. If this e-mail travels at the speed of light, it will take 100 years to reach us, so if we receive it today it was sent 100 years ago. If we send a reply it will take another 100 years to reach them. There will be a gap of 200 years between when they send the message and receive our reply. Our position, here and now, cannot be connected to their position during that 200-year period without travelling faster than light. The square of the 'time-distance' between one and the other is positive.

When the square of the 'time-distance' between two points – or

I should say two *events* – in space-time is zero, then the events can be connected by a light ray, emitted from one and received by the other. A light ray experiences no time; it is as if it were travelling instantaneously from one point to another, though the mathematics allows us to measure a finite speed. At the speed of light time stands still, and it makes no sense to travel faster than light, unless you can move backwards in time.

Einstein's *special* theory of relativity gave way ten years later to his *general* theory of relativity, in which space-time is curved in order to incorporate gravity. This works well at the macroscopic level, except in a black hole, where the mass of an object is too great for the space it occupies, and the enormous curvature leads to a singularity in space-time.

At the microscopic level – at the level of atoms and molecules – gravitational forces are so low as to be insignificant. Physicists could ignore gravity as they started examining the internal structure of atoms, but they had to develop quantum theory instead. In an atom they found that most of the mass was concentrated in a tiny nucleus, and this was composed of particles called protons and neutrons. Further investigation showed that protons and neutrons also had an internal structure, involving quarks. However, the process of finding internal structures could not go on indefinitely because if the mass became concentrated in ever smaller 'particles' they would eventually turn into black holes. At very high energies quantum mechanics and general relativity are inconsistent, but in the 1970s a new theory began to emerge. This was string theory, where the particles were viewed as strings moving through space-time.

Physicists see string theory as *the* way to bring quantum mechanics and general relativity into agreement. It changes quantum mechanics, but general relativity has to change too. The really big change is that there have to be further dimensions to space-time. Four dimensions are not sufficient, and the minimum number of dimensions is ten. The extra dimensions are tightly curled up on themselves, like the surfaces of tiny tubes, which at the usual macroscopic level are invisible. String theory is an attempt to unify relativity theory and quantum theory, giving a quantum structure to space-time.

The idea of going beyond four dimensions appeals to mathematicians, and has appeared frequently in this book. We have thought of it as expanding three-dimensional Euclidean geometry by adding extra dimensions, but in this case we are expanding four-dimensional Minkowski geometry by adding extra dimensions. There is an important difference. The distance between two points depends on the differences in their coordinates. In Euclidean space the square of the distance is the sum of the squares of these coordinates. But in Minkowski geometry the 'sum' involves exactly one minus sign. When we expand to higher dimensional space-time we want to retain this minus sign, and such a space is called Lorentzian.* General relativity uses a curved version of Minkowski geometry, and string theory uses a curved version of Lorentzian geometry.

The number of dimensions for string theory seems to be 10 or 26, and 26 is particularly intriguing, for the following reason. A light ray in Lorentzian space – meaning a path on which the 'time-distance' is always zero – yields a 'perpendicular' Euclidean space of two dimensions lower. Doing this with 26-dimensional Lorentzian space yields 24-dimensional Euclidean space, which is where the Leech Lattice lives.

This is more than a coincidence of dimensions, because it turns out that 26-dimensional Lorentzian space contains a remarkable lattice that is unique in an important technical sense. Choosing a light ray in this lattice yields a lattice in 24-dimensional Euclidean space, and depending on the light ray you take, any one of 24 possible lattices arises. One of these is the Leech Lattice.

Before I explain how to find a light ray that will give the Leech Lattice, here is a remarkable fact:

$$1^2 + 2^2 + 3^2 + 4^2 + \ldots + 21^2 + 22^2 + 23^2 + 24^2 = 70^2$$

The sum of the first 24 squares is a square! This is phenomenal. Twenty-four is the only whole number larger than 1 for which it happens. The sum of the first n squares is never a perfect square otherwise.

Now back to 26-dimensional Lorentz space. The exceptional lattice that I mentioned contains a point whose coordinates are:

$$(0,1,2,3,4,5,6,7,8,9,10,11,12,13,14,15,16,17,18,19,20,21,22,23,24,70)$$

This point lies on a light ray through the origin – through the point whose coordinates are all 0 – because the 'time-distance' from the origin to this point is:

$$0^2 + 1^2 + 2^2 + 3^2 + 4^2 + \ldots + 21^2 + 22^2 + 23^2 + 24^2 - 70^2 = 0$$

This light ray yields the Leech Lattice. If the Pythagoreans were still with us today they might see this as evidence that the universe is indeed based on whole numbers!

In 1990, Borcherds used the crystalline structure of the 26-dimensional Lorentzian Lattice, rather than the Leech Lattice itself, in creating his fake Monster Lie algebra. He was now getting very close to a proof of the Conway–Norton conjectures, and two years later, in 1992, he published a paper having the title 'Monstrous Moonshine and Monstrous Lie Algebras'. Here he uses the work of Frenkel, Lepowsky and Meurman to create a new Monster Lie algebra, which he then applies to prove that the moonshine module satisfies the Conway–Norton Moonshine conjectures.

Borcherds's work was moving closely towards mathematical physics, and two years later he followed up with a paper where he creates an algebra structure by quantizing a string moving in space-time, showing that 'It turns out to be non-zero only if space-time is 26-dimensional.'* If string theory needs 26 dimensions, as opposed to ten, then perhaps the 1980s quotation by Freeman Dyson in the Prologue will turn out to be quite prescient. The Monster may indeed be built into the structure of the universe.

In 1998, Borcherds won the greatest prize in mathematics for his work: the Fields Medal. To win a Fields Medal a mathematician has to be under 40 years old, and the award takes place at the International Congress of Mathematicians, which is held every four years – in 1998 it was in Berlin. The work leading to the medal is described by an eminent and more senior mathematician, and in Borcherds's case it was Peter Goddard, a mathematical physicist from Cambridge (now head of the Institute for Advanced Study in Princeton). In his illuminating description of Borcherds's work, Goddard ends by saying:

> Displaying penetrating insight, formidable technique and brilliant originality, Richard Borcherds has used the beautiful properties of some exceptional structures to motivate new algebraic theories of great power with profound connections with other areas of math-

ematics and physics. He has used them to establish outstanding conjectures and to find new deep results in classical areas of mathematics. This is surely just the beginning of what we have to learn from what he has created.*

The Fields Medal appeared earlier, when it was awarded to John Thompson in 1970. It is a rarer honour than the Nobel Prize, but less well known, and some mathematicians regret the lack of a Nobel Prize in mathematics. This leads to a perennial story about Nobel's wife having had an affair with a mathematician, but the story is certainly false: the mathematician lived in Norway, and Nobel, despite his Swedish nationality, lived in Paris; moreover, Nobel was a confirmed bachelor. In 1985 two Swedish mathematicians, Lars Gårding and Lars Hörmander published an article on this topic in the *Mathematical Intelligencer* in which they conclude that 'mathematics was simply not one of Nobel's interests'.*

In recent years, however, the Norwegian government has moved to rectify the situation. In 2002, on the 200th anniversary of Abel's birth (Niels Henrik Abel appeared in Chapter 2), they established a fund to support an Abel prize in mathematics. It is intended to be similar to the Nobel prize, and the first award was in 2003 to Jean-Pierre Serre, who appeared briefly in Chapter 15.

Mathematicians toil in obscurity because the subject is a difficult one to explain and has moved an enormous way in the past four thousand years since the Babylonians solved quadratic equations. Barring human catastrophe it will continue for thousands of years more, and there will still be unsolved problems, and mysteries to inspire further work. The Moonshine mystery itself is still unresolved, despite Borcherds's proof!

Borcherds used the replication formulas of Conway and Norton to reduce more than a hundred cases down to just four, which he then proved. This was a great piece of work, but as Conway says, 'His real achievement was to put it all in a big theory, although it still doesn't give a conceptual explanation.' In mathematics we prove theorems, but we also want to *understand* things, and there are facts about the Monster and Moonshine that we don't understand. Here is one.

When Conway and Norton were working on Moonshine they used the columns of the Monster's character table to obtain mini-*j*-functions, as I mentioned in Chapter 15. There are 194 columns, but for elementary reasons some give the same function. This reduces the number of functions to 171, but Conway and Norton wanted to know how many of these functions were completely independent of one another in the sense that you couldn't get one of them by adding and subtracting others. They started looking at dependencies between the different functions, and found themselves gradually counting down from 171. Conway recalls that, 'As we went down into the 160s, I said let's guess what number we will reach.' They guessed it would be 163 – which has a very special property in number theory – and it was!

There is no explanation for this. We don't know whether it is merely a coincidence, or something more. The special property of 163 in number theory has intriguing consequences, among which is the fact that

$$e^{\pi\sqrt{163}} = 262537412640768743.99999999999925\ldots$$

is very close to being a whole number. Here π is the famous ratio of the circumference of a circle to its diameter, and e is almost as famous, being the basis for natural logarithms and exponential growth. The close approximation to a whole number is no mere chance. It uses the *j*-function, and a special feature of the number 163*.

When McKay made his observation about 196,883 and 196,884 appearing in the Monster and the *j*-function, the numbers were high enough to make a striking coincidence, but 163 is low by comparison, and it is difficult to say whether it will lead anywhere. McKay himself noticed a very strange correspondence between one class of mirror symmetries in the Monster and the Lie group of type *E*8 (Chapter 5), and although it involves a pattern of much smaller numbers, from 1 to 6, some very recent work shows that it is no chimera. Mathematicians in Japan and Taiwan show that vertex algebras offer a foundation for this connection.* McKay also points out that the number of dimensions for the string theory associated with the Monster – namely 26 – is the same as the number of *types* of operations in Mathieu's group *M*24, which is the first step on the road to the Monster. Perhaps it is only a coincidence that there are so many coincidences, but we do not know.

Strange connections like these were not the reason mathematicians discovered the Monster, but a consequence. The Monster was revealed as the result of a long process, starting with Galois's work in about 1830. It was he who discovered that there are 'simple' groups of permutations – groups that cannot be deconstructed – which I have dubbed here atoms of symmetry, and many more were found later. By the early 1960s there was an extensive table, along with five exceptions from the nineteenth century. The Feit–Thompson theorem in 1963 then made it feasible to find and classify any others, leading to the great Classification project. Three years later, with Janko's sudden publication of a new exception, there was an enormous impetus to find others, and during the next ten years another 20 were discovered, bringing the total to 26. The second largest, called the Baby Monster, was found in an inspired search by Fischer, and from the Baby emerged the Monster itself, the largest of the exceptions.

The method leading to its discovery, brilliant though it was, gave no clue to the Monster's remarkable properties. It was only later that the first hints arose of odd coincidences between the Monster and number theory, and these were to lead to the connection with string theory. The Moonshine connections between the Monster and number theory have now been placed within a larger theory, but we have yet to grasp the significance of these deep mathematical links with fundamental physics. We have found the Monster, but it remains an enigma. Understanding its full nature is likely to shed light on the very fabric of the universe. But that story must await a future book.

Notes

Prologue

3 Goethe, Naturwissenschaftliche Schriften, in *Collected Works* (author's translation).

4 F. J. Dyson, Unfashionable pursuits, *Mathematical Intelligencer*, 5 (1983), 47–54.

Chapter 2: Galois: Death of a Genius

12 Laura Toti-Rigatelli, *Evariste Galois*, English translation, Birkhäuser, 1996, p. 113.

18 John Fauvel and Jeremy Gray, *A History of Mathematics*. The Open University, 1987, p. 255.

19 See http://turnbull.mcs.st-and.ac.uk/~history/Mathematicians/

George Sarton, *Six Wings: Men of Science in the Renaissance*, Indiana University Press, 1957, p. 28.

Jean-Pierre Tignol, *Galois' Theory of Algebraic Equtions*, English translation, Longman, 1988, p. 274.

20 Tignol, *Galois' Theory*, p. 274, translation altered slightly.

24 Toti-Rigatelli, *Evariste Galois*, p. 98.

25 Toti-Rigatelli, *Evariste Galois*.

Chapter 3: Irrational Solutions

33 Fauvel and Gray, *A History of Mathematics*, pp. 504, 505.

35 Fauvel and Gray, *A History of Mathematics*, p. 503.

Fauvel and Gray, *A History of Mathematics*, p. 503.

Chapter 4: Groups

45 Quoted in Yu. I. Manin, *Mathematics and Physics*, Birkhäuser, 1981, p. 35.

46 A proof of Fermat's Last Theorem was finally given by Andrew Wiles in the mid-1990s.

Chapter 5: Sophus Lie

53 A. Stubhaug, *The Mathematician Sophus Lie*, English translation, Springer, 2002, p. 3.

54 Stubhaug, *The Mathematician Sophus Lie*, p. 9.

55 Stubhaug, *The Mathematician Sophus Lie*, p. 10.

56 Stubhaug, *The Mathematician Sophus Lie*, p. 12.

63 There are three degrees of freedom for the velocity (direction of motion and speed), and three for the spin (direction and amount of spin).

This quote and all further extracts from correspondence in this chapter are taken from T. Hawkins, *Emergence of the Theory of Lie Groups*, Springer, 2000.

Chapter 6: Lie Groups and Physics

72 Stubhaug, *The Mathematician Sophus Lie*, p. 377.

Stubhaug, *The Mathematician Sophus Lie*, p. 376.

73 Stubhaug, *The Mathematician Sophus Lie*, p. 375.

75 Quotes by Bohr and Feynman at http://turnbull.mcs.st-and.ac.uk/ ~history/Quotations/index.html

76 Any particle carrying electric charge is susceptible to the electro-magnetic force, but the force itself is mediated by photons, which have no electrical charge. The gauge group is called U(1).

These are called bosons of types W^+, W^-, and Z. The gauge group for the weak nuclear force is called SU(2).

77 The gauge group for the strong nuclear force is called SU(3).

Chapter 7: Going Finite

82 You can even use powers of prime numbers: $4 = 2^2$, $8 = 2^3$, $9 = 3^2$, and so on. But beware, this is no longer cyclic arithmetic – it's more complicated than that.

85 W. Burnside, *Theory of Groups of Finite Order*, Cambridge University Press, 1897.

86 Peter Neumann, The context of Burnside's contributions to group theory, in Peter M. Neumann, A. J. S. Mann and, Julia C. Tompson (eds), *The Collected Papers of William Burnside*, two volumes, Oxford University Press, 2004.

Chapter 8: After the War

90 N. Bourbaki, Foundations of mathematics for the working mathem-atician, *Journal of Symbolic Logic*, 14 (1949), 1–8.

91 A. Borel, Twenty-five years with Nicolas Bourbaki (1949–1973), *Notices of the American Mathematical Society*, 45 (1998), 373–80.

Interview with Henri Cartan, *Notices of the American Mathematical Society*, 46 (1999), 782–8.

92 Borel, Twenty-five years with Nicolas Bourbaki.

93 J. Dieudonné, The work of Nicolas Bourbaki, *American Mathematical Monthly*, 77 (1970), 135–45.

http://turnbull.mcs.st-and.ac.uk/~history/Quotations/index.html

Chapter 9: The Man from Uccle

108 We called this genetic code a 'blueprint' because of the metaphorical connection with buildings, but 'genetic code' sounds better in the context of growing a multi-crystal.

110 A block system like this is the same as a multi-crystal based on hexagons, the block size being the number of edges per vertex.

Chapter 10: The Big Theorem

116 S. Mac Lane, Mathematics at Göttingen under the Nazis, *Notices of the American Mathematical Society*, 42 (1995), 1134–8.

M. Schiffer, Issai Schur: some personal reminiscences, in H. Begehr (ed.), *Mathematik in Berlin: Geschichte und Dokumentation*, Aachen, 1998.

W. Feit, Richard D. Brauer, *Bulletin of the American Mathematical Society*, 10 (1978), 1–20.

117 W. Feit, Richard D. Brauer.

J. A. Green, Richard Dagobert Brauer, *Bulletin of the London Mathematics Society*, 10 (1978), 317–42.

118 A cross-section is called an 'involution centralizer' in mathematical terms: 'involution' refers to a symmetry of order 2, and 'centralizer'

refers to the sub-group that leaves the involution alone – for example, by keeping the mirror in place.

119 D. Gorenstein, *Finite Simple Groups: An Introduction to their Classification*, Plenum Press, 1982, p. 1.

120 www.math.yale.edu/public_html/WalterFeit/ToNewYork/Welcome.html

123 D. Gorenstein, *Finite Simple Groups*, p. 16.

124 Richard Brauer, On the work of John Thompson, *Proceedings of the International Congress of Mathematicians*, 1 (1970), 15–16; also in *Richard Brauer: Collected Papers Volume III*, MIT Press, 1980, pp. 688–9.

Chapter 11: Pandora's Box

130 A permutation group on five or more objects exhibiting at least 3-fold transivity has to involve a symmetry atom.

133 I. Kersten, Biography of Ernst Witt (1911–1991), *Contemporary Mathematics*, 272 (2000), 155–71.

134 The old Egyptian language continued to be used, but people started writing it using the Greek alphabet, plus a few extra letters. The understanding of hieroglyphs was lost, and they were only deciphered in the nineteenth century.

136 Gorenstein, *Finite Simple Groups*, p. 85.

138 And *J*3 was later constructed on a computer by Graham Higman at Oxford and John McKay at Montreal; many years later it was also constructed by hand as a group of permutations by Richard Weiss in Boston.

139 Gerhard Hiss, Die sporadischen Gruppen, *Jahresbericht der Deutschen Mathematiker-Vereinigung*, 105 (2003), 169–94.

Chapter 12: The Leech Lattice

148 H. Cohn and H. Kumar, The densest lattice in twenty-four dimensions, *Electronic Research Anouncements of the American Mathematical Society*, 2004 (www.mpim-bonn.mpg.de/external-documentation/era-mirror/era-msc–2004.html).

149 Donald Higman and Graham Higman are not related. They just happened to work in the same area of mathematics.

150 Quotations from Conway appear in Thomas Thompson, *From Error-correcting Codes through Sphere Packings to Simple Groups*, Carus Mathematical Monograph 21, Mathematical Association of America, 1983.

156 John Conway, *On Numbers and Games*, Academic Press, 1976; Elwyn Berlekamp, John Conway, and Richard Guy, *Winning Ways for Your Mathematical Plays*, Academic Press, 1982.

Courtesy of W. O. J. Moser.

Chapter 13: Fischer's Monsters

161 To be precise he called them '3-transpositions'.

Using 2-cyclic, or in one case 3-cyclic, arithmetic.

Chapter 14: The Atlas

173 R. Solomon, On finite simple groups and their classification, *Notices of the American Mathematical Society*, 42 (1995), 231–9.

174 1989 Steele Prizes, *Notices of the American Mathematical Society*, 36 (1989), 831–6.

175 1989 Steele Prizes.

181 One operation in the group does nothing; six operations switch a pair of beads and leave the other two alone; three operations switch

two pairs of beads simultaneously; eight operations fix one bead and move the other three cyclically among themselves; and finally there are six operations moving all four beads cyclically among themselves.

Chapter 15: A Monstrous Mystery

195 Dictionary of Scientific Biography: Bolyai.

196 Dictionary of Scientific Biography: Bolyai.

199 The correct mathematical term for mini–*j*–function is *Hauptmodul*.

200 J. H. Conway, Monsters and Moonshine, *Mathematical Intelligencer*, 2 (1980), 165–71.

202 Letter from J. Thompson to P. Fong, 19 March 1979.

Chapter 16: Construction

205 The simplest way to multiply two *n* by *n* matrices involves n^3 (*n* cubed) multiplications of pairs of numbers – if *n* is 200,000, then n^3 is 8,000,000,000,000,000. Doing five billion of these a second would take about six months.

210 J. H. Conway, A simple construction for the Fischer–Griess monster group, *Inventiones Mathematicae*, 79 (1985), 513–40.

211 Conway, Monsters and Moonshine.

212 Solomon, On finite simple groups and their classification.

213 Conway, Monsters and Moonshine.

Chapter 17: Moonshine

218 Igor Frenkel, James Lepowsky and Arne Meurman, *Vertex Operator Algebras and the Monster*, Academic Press, 1988.

220 Borcherds based his ideas on earlier work of Victor Kac and Robert Moody, who created infinite-dimensional Lie algebras, analogous to the finite-dimensional ones used by Killing and Cartan.

223 This term is in honour of Hendrik Lorentz (1853–1928), a Dutch physicist who was an early pioneer of relativity theory. He formulated the changes we would observe in objects moving at speeds close to the speed of light. These became known as the Lorentz transformations, and were incorporated in a natural way in Minkowski's geometry of four-dimensional space-time.

225 R. Borcherds, Sporadic groups and string theory, *First European Congress of Mathematics*, 1 (1994), 411–21.

226 P. Goddard, The work of Richard Ewen Borcherds, *Documenta Mathematica*, extra volume, 1 (1998), 99–108.

L. Gårding and L. Hörmander, Why is there no Nobel Prize in mathematics?, *Mathematical Intelligencer*, 7 (1985), 73–4.

228 This special feature also yields a fact, first noticed by Euler, that the formula $x^2 - x + 41$ gives prime numbers for all values of x between 1 and 40. The connection with 163 is that the solution to $x^2 - x + 41 = 0$ involves the square root of -163.

228 C. H. Lam, H. Yamada and H. Yamauchi, Vertex operator algebras, extended *E8* diagram, and McKay's observation on the Monster simple group, http://arxiv.org/abs/math.QA/0403010; and http://arxiv.org/abs/math.QA/0503239

Appendix 1

The Golden Section

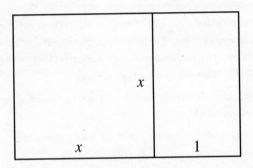

In this picture the small rectangle and the large rectangle have the same proportions, meaning the same ratio of length to width. This ratio is the golden section. In the small rectangle it is $x/1$, and in the large rectangle it is $(x + 1)/x$. This gives the equation:

$$x/1 = (x + 1)/x$$

Multiplying both sides by x yields:

$$x^2 = x + 1$$

This is a quadratic equation. When we write it in the alternative form as

$$x^2 - x - 1 = 0$$

and use the quadratic formula, we find its two solutions are:

$$x = \frac{1 + \sqrt{5}}{2} \quad \text{and} \quad x = \frac{1 - \sqrt{5}}{2}$$

The one with a plus sign is the golden section. It works out at approximately 1.618 . . .

Appendix 2

The Witt Design

Mathieu's group $M24$ permutes 24 symbols in such a way that any sequence of five symbols can be sent to any other. No other group – except Mathieu's group $M12$, which permutes 12 symbols – can do this unless it contains all even permutations.

In 1934–5, Ernst Witt constructed a remarkable design using 24 symbols, and having $M24$ as its group of symmetries. Witt's design is a collection of subsets, called octads, each having eight symbols, and with the property that each set of five symbols lies in exactly one octad. The number of octads must be 759, as I shall now demonstrate.

First count the number of sequences of five symbols. We are choosing from 24 symbols, so there are 24 choices for the first symbol, 23 for the second one, 22 for the third, 21 for the fourth, and 20 for the fifth. The number of such quintuples is therefore:

$$24 \times 23 \times 22 \times 21 \times 20$$

Now count them in a different way. The number of quintuples in each octad is $8 \times 7 \times 6 \times 5 \times 4$ (eight choices for the first member of the quintuple, seven for the second, etc.). Each quintuple lies in exactly one octad, so if N denotes the number of octads, then the number of quintuples must be $N \times 8 \times 7 \times 6 \times 5 \times 4$. Hence:

APPENDIX 2: THE WITT DESIGN

$$N \times 8 \times 7 \times 6 \times 5 \times 4 = 24 \times 23 \times 22 \times 21 \times 20$$

This gives

$$N = \frac{24 \times 23 \times 22 \times 21 \times 20}{8 \times 7 \times 6 \times 5 \times 4} = 759$$

Appendix 3

The Leech Lattice

The Leech Lattice gives the tightest lattice packing of spheres in 24 dimensions. The points of the lattice are the centres of spheres, each of which touches 196,560 others – this is the maximum possible in 24 dimensions. Each lattice point can be specified using 24 coordinates labelled by the 24 symbols of Witt's design. Take one sphere centred at the origin, so the coordinates of that point are all zero. The centres of the 196,560 neighbouring spheres split naturally into three sub-sets of sizes

$$97,152 + 1,104 + 97,308 = 196,560.$$

The sub-set of size 97,152

This number is $2^7 \times 759$. There are 759 octads in Witt's design (see Appendix 2), and for each one there are 2^7 points. The coordinates of each point are ± 2 in the positions of an octad, and zero elsewhere; the number of minus signs is even.

The sub-set of size 1,104

This number is $2^2 \times 276$. There are 276 ways of choosing two coordinates from 24: each of these two coordinates is ± 4, and the other 22 coordinates are zero.

The sub-set of size 98,304

This number is $2^{12} \times 24$. One coordinate is ± 3, the others are ± 1. Witt's design provides a way of making 2^{12} sign choices.

The distance of a point from the origin, when squared, is the sum of the squares of its coordinates – this is Pythagoras's theorem generalized to n dimensions. The sums of the squares of the coordinates for the 196,560 points are all equal.

In the first sub-set: $\quad 2^2 + 2^2 + 2^2 + 2^2 + 2^2 + 2^2 + 2^2 + 2^2 = 32$

In the second sub-set: $\qquad\qquad\qquad\qquad 4^2 + 4^2 = 32$

In the third sub-set: $\qquad\qquad\quad 3^2 + 1^2 + 1^2 + \ldots + 1^2 = 32$

This shows that all these 196,560 points are an equal distance from the origin.

Appendix 4

The 26 Exceptions

Here is a list of the 26 exceptional symmetry atoms – the so-called sporadic groups.

Name	Symbol	Size
Mathieu groups	$M11$	$7920 = 2^4 \cdot 3^2 \cdot 5 \cdot 11$
	$M12$	$95{,}040 = 2^6 \cdot 3^3 \cdot 5 \cdot 11$
	$M22$	$443{,}520 = 2^7 \cdot 3^2 \cdot 5 \cdot 7 \cdot 11$
	$M23$	$10{,}200{,}960 = 2^7 \cdot 3^2 \cdot 5 \cdot 7 \cdot 11 \cdot 23$
	$M24$	$244{,}823{,}040 = 2^{10} \cdot 3^3 \cdot 5 \cdot 7 \cdot 11 \cdot 23$
Janko groups	$J1$	$175{,}560 = 2^3 \cdot 3 \cdot 5 \cdot 7 \cdot 11 \cdot 19$
	$J2$	$604{,}800 = 2^7 \cdot 3^3 \cdot 5^2 \cdot 7$
	$J3$	$50{,}232{,}960 = 2^7 \cdot 3^5 \cdot 5 \cdot 17 \cdot 19$
	$J4$	$86{,}775{,}571{,}046{,}077{,}562{,}880 =$ $2^{21} \cdot 3^3 \cdot 5 \cdot 7 \cdot 11^3 \cdot 23 \cdot 29 \cdot 31 \cdot 37 \cdot 43$
Higman–Sims	HS	$44{,}352{,}000 = 2^9 \cdot 3^2 \cdot 5^3 \cdot 7 \cdot 11$
McLaughlin	Mc	$898{,}128{,}000 = 2^7 \cdot 3^6 \cdot 5^3 \cdot 7 \cdot 11$
Held	He	$4{,}030{,}387{,}200 = 2^{10} \cdot 3^3 \cdot 5^2 \cdot 7^3 \cdot 17$

Name	Symbol	Size
Suzuki	*Suz*	$448,345,497,600 =$ $2^{13} \cdot 3^7 \cdot 5^2 \cdot 7 \cdot 11 \cdot 13$
Rudvalis	*Ru*	$145,926,144,000 =$ $2^{14} \cdot 3^3 \cdot 5^3 \cdot 7 \cdot 13 \cdot 29$
O'Nan	*ON*	$460,815,505,920 =$ $2^9 \cdot 3^4 \cdot 5 \cdot 7^3 \cdot 11 \cdot 19 \cdot 31$
Lyons	*Ly*	$51,765,179,004,000,000 =$ $2^8 \cdot 3^7 \cdot 5^6 \cdot 7 \cdot 11 \cdot 31 \cdot 37 \cdot 67$
Conway groups	*Co*1	$4,157,776,806,543,360,000 =$ $2^{21} \cdot 3^9 \cdot 5^4 \cdot 7^2 \cdot 11 \cdot 13 \cdot 23$
	*Co*2	$42,305,421,312,000 =$ $2^{18} \cdot 3^6 \cdot 5^3 \cdot 7 \cdot 11 \cdot 23$
	*Co*3	$495,766,656,000 =$ $2^{10} \cdot 3^7 \cdot 5^3 \cdot 7 \cdot 11 \cdot 23$
Fischer groups	*Fi*22	$64,561,751,654,400 =$ $2^{17} \cdot 3^9 \cdot 5^2 \cdot 7 \cdot 11 \cdot 13$
	*Fi*23	$4,089,470,473,293,004,800 =$ $2^{18} \cdot 3^{13} \cdot 5^2 \cdot 7 \cdot 11 \cdot 13 \cdot 17 \cdot 23$
	*Fi*24	$1,255,205,709,190,661,$ $721,292,800 =$ $2^{21} \cdot 3^{16} \cdot 5^2 \cdot 7^3 \cdot 11 \cdot 13 \cdot 23 \cdot 29$
Harada–Norton	*HN*	$273,030,912,000,000 =$ $2^{14} \cdot 3^6 \cdot 5^6 \cdot 7 \cdot 11 \cdot 19$

Thompson	*Th*	$90,745,943,887,872,000 =$ $2^{15} \cdot 3^{10} \cdot 5^3 \cdot 7^2 \cdot 13 \cdot 19 \cdot 31$
Baby Monster	B	$4,154,781,481,226,426,191,$ $177,580,544,000,000 =$ $2^{41} \cdot 3^{13} \cdot 5^6 \cdot 7^2 \cdot 11 \cdot 13 \cdot 17 \cdot 19 \cdot 23 \cdot$ $31 \cdot 47$
Monster	M	$808,017,424,794,512,875,$ $886,459,904,961,710,757,005,$ $754,368,000,000,000 =$ $2^{46} \cdot 3^{20} \cdot 5^9 \cdot 7^6 \cdot 11^2 \cdot 13^3 \cdot 17 \cdot 19 \cdot 23 \cdot$ $29 \cdot 31 \cdot 41 \cdot 47 \cdot 59 \cdot 71$

This table shows immediately that some groups cannot possibly be sub-groups of others, because Lagrange's theorem says the size of a group must divide the size of any larger group containing it. Writing each size as a product of prime numbers makes this easy to check. For example, the size of $M12$ is a multiple of 3^3 but the size of $M22$ is not a multiple of 3^3 so $M12$ can't be a sub-group of $M22$. A similar argument shows that neither the Lyons group nor the fourth Janko group $J4$ can be sub-groups of the Monster: the size of each one is divisible by 37, but the size of the Monster is not divisible by 37. More technical arguments show that $J1$, $J3$, Ru, and ON cannot be involved in the Monster. A detailed plan of how these 26 exceptions are involved with one another is given on the next page.

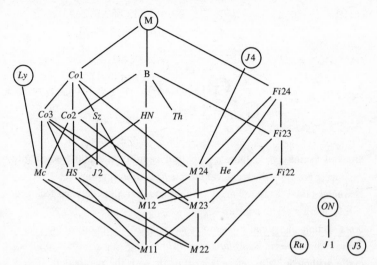

The network above shows the involvement of one exceptional symmetry atom in another. The circled ones are involved in nothing larger. The Monster involves all but six of the other exceptions, omitting only *J*4, *Ly, ON, Ru, J*1, and *J*3.

Glossary

atom of symmetry   a finite group that cannot be deconstructed into simpler groups – the technical term is *finite simple group*.

character table   a square array of numbers that gives detailed technical information about a group.

cross-section   a special sub-group associated to an operation of order 2; the technical term is *involution centralizer*.

cyclic arithmetic   this refers to arithmetic with the numbers 0, 1, 2, 3, . . ., n, where n is the same as 0. The technical term is *modular arithmetic*.

deconstruction   this refers to deconstructing a group into a series of layers, each of which is a 'simple' group. Technically, these layers form a *composition series*, and deconstruction is called *decomposition*.

group   a group can be seen as a system of operations, each of which is reversible, and in which one operation followed by another is a third operation in the same system.

group, cyclic   a group generated by a single operation. For example, a rotation by 60° generates a cyclic group of size 6 containing rotations by angles 60°, 120°, 180°, 240°, 300°, and 0°.

group, prime cyclic   a cyclic group whose size is a prime number.

group, simple   a group that cannot be deconstructed; what I have called an atom of symmetry.

group, sporadic   one of the 26 exceptional symmetry atoms.

hyperbolic plane   a plane geometry in which Euclid's parallel axiom fails, and the angles of a triangle add up to less than 180°.

GLOSSARY

***j*-function**   this assigns a number to each point of the hyperbolic plane, and is closely associated to the modular group.

Leech Lattice   a remarkable lattice giving the tightest packing of spheres in 24 dimensions.

Lie group   a group in which the operations can be continuously modified.

mini-*j*-function   like the *j*-function, but associated with a group derived from the modular group. The technical term is *Hauptmodul*.

modular group   a group derived from the symmetry group for the hyperbolic plane obtained by restricting from the real numbers to the integers.

periodic table   a table of finite symmetry atoms, in seven families *A–G*, that includes all but the 26 exceptions. The technical term is the collection of *groups of Lie type*.

permutation   the process of rearranging a collection of objects.

Witt design   a remarkable design on 24 symbols that is used in constructing Mathieu's group *M*24, and the Leech Lattice.

Index

Abel prize, 226
Abel, Niels Henrik, 20, 226
Alperin, Jonathan, 123, 125
Aschbacher, Michael, 165–8, 174,
    214
Atkin, A. O. L., 191, 201–2
Atlas, 179–85
Atlas Laboratory (computer
    centre), 149, 190
atom of symmetry, 8, 40
    exceptional, 128, 155, 161, 179,
    211, 244–7

Baby Monster, 159, 165, 178–80
Babylonians, 16
Baer, Reinhold, 132, 158–9
black hole, 222
Bletchley Park, 202
block system, 109–11
Bolyai, Farkas, 195–6
Bolyai, Janos, 195–6
Borcherds, Richard, 4, 219–20,
    224–5
Borel, Armand, 91
Bourbaki, Nicolas, 90–4, 112

Brauer, Alfred, 115
Brauer, Richard, 115–9, 121, 124,
    202
building 98, 111–12, *see also*
    multi-crystal
Burnside, theorem of, 86
Burnside, William, 84–7, 123

calculus, 87, 89
Cameron, Peter, 134
Cardano, Gerolamo, 17–19
Cartan, Élie, 69–70, 73
Cartan, Henri, 91–2
Cauchy, Augustin, 14, 26,
    46–7
character table, 134, 181
character theory, 86
Chevalley, Claude, 94–5, 99
Classification (of finite simple
    groups), 119, 136, 171–5,
    212–14
Conway groups, 155
Conway, John Horton, 2, 148–56,
    178–83, 199–203, 210–11,
    213, 227

Conway–Norton conjectures, 217, 219, 225
coordinate geometry, 60
cross-section, 118–19
cube, 5–9, 103–15
cube, symmetry of, 50
Curtis, Robert, 179–82
cyclic arithmetic, 80–2, 83–4

decomposition, *see* deconstruction
deconstruction, 34
del Ferro, Scipione, 17–19
Deligne, Pierre, 108
Descartes, René, 59
Dickson, Leonard Eugene, 79–80, 82–3, 88
Dieudonné, Jean, 92
dodecahedron, 6, 9, 103–4
Dürer, Albrecht, 28
Dyson, Freeman, 4

Einstein, Albert, 72–3, 220–2
electromagnetism, 72
electron, 75
electron orbital, 75–6
Engel, Friedrich, 64, 68, 71
equation, algebraic, 15–21
    differential, 57–8
    irreducible, 27, 29, 32
Euclid, 27, 90, 196–7
Euler, Leonard, 33, 202

Feit, Walter, 114, 119–23, 164, 168
Feit-Thompson theorem, 117, 119
Fermat's Last Theorem, 46

Ferrari, Ludovico, 17–19
Feynman, Richard, 45, 75
Fibonacci, Leonardo, 28–9
Fields medal, 124, 225
Fischer groups, 161–4, 178
Fischer, Bernd, 157–9, 160–1, 175–88, 191–3
Fong, Paul, 202
Fontana, Niccolo, *see* Tartaglia
four dimensions, 60–1
Fourier, Joseph, 22
Frenkel, Igor, 218

Galois group, 34
Galois, Évariste, 11–15, 20–6, 31–5
Gauss, Carl Friedrich, 11, 19
Goddard, Peter, 225
Goethe, Johann Wolfgang von, 3, 10
Goldbach conjecture, 113
golden section, 27–30, 238–9
Gorenstein, Daniel, 119, 123, 135, 172–5, 211, 213
Gosset, Thorold, 106
gravitation, 73, 87
Griess, Robert, 178, 206–10
group, 33
group of Lie type, *see* periodic table
group, as abstract concept, 50–1
    cyclic, 34
    of permutations, 42–3
    prime cyclic, 34–5
    simple, *see* atom of symmetry

sporadic, *see* atom of symmetry, exceptional
sub-group, 43, 45

Hales, Thomas, 146
Hall, Marshall, 136, 138–40
Harada, Koichiro, 184
Held, Dieter, 137–8, 179
Higman, Donald, 139
Higman, Graham, 149
Hilbert, David, 77, 131
Hughes, Daniel, 122
hyperbolic plane, 196

icosahedron, 6, 9, 103–4
interchange, *see* permutation
involution centralizer, *see* cross-section
involution, *see* mirror symmetry, transposition

Janko groups, 135–6, 137–8, 189
Janko, Zvonimir, 125–8, 134–8, 179, 189, 214
j-function, 191–4, 199, 217–19, 227–8
j-function, mini-, 199, 217
Jordan, Camille, 48–9, 82

Kepler conjecture, 146–7
Khayyám, Omar, 16–17
Killing, Wilhelm, 63–8
Killing-Cartan classification, 70
Klein, Felix, 56, 63–4

Lagrange, Joseph-Louis, 19, 43–4
Lagrange, theorem of, 43, 45
Leech Lattice, 147–55, 179, 216–17, 220, 242–3
Leech, John, 147–9
Leon, Jeffrey, 188
Leonardo da Vinci, 28
Lepowsky, James, 218
Lie algebras, 220
Lie groups, 59, 62, 76, 79
    classical, 66, 83
Lie theory, 72
Lie, Sophus, 53–59, 62–5, 67–8, 69, 70
Livingstone, Donald, 185–8
Lobachevski, Nikolai, 195–6
Lorentzian, 223–4
Louis-Phillipe, King of France, 23
Loyd, Sam, 36–8
Lyons, Richard, 179, 211, 213

Mac Lane, Saunders, 115–16, 121, 173
Mason, Geoffrey, 212
Mathieu groups, 131, 133, 161–3
Mathieu, Émile, 130–1, 133
McKay, John, 1, 149, 190, 228
McLaughlin, Jack, 155
Meurman, Arne, 218
Miller, G. A., 131
Minkowski geometry, 221, 223
Minkowski, Hermann, 73, 221
mirror symmetry, abstract, 162–5
    as reflection, 7, 162

modular arithmetic, *see* cyclic arithmetic

modular group, 198

Monster, 119, 133, 178–9, 190–3, 199, 216–17, 227–9

Monster, construction of, 205–11

Monster Lie algebra, 225

Monster Lie algebra, fake, 220, 224

moonshine, 203, 216–29

moonshine module, 219–20, 225

multi-crystal, 99–104, 107, 109

multidimensional crystal, 104

multidimensional space, 61, 74

Napoleon, 12–13

Nazi party, 115–17, 132

Neumann, Peter, 152

Newton, Isaac, 87

Nobel prize, 124, 226

Noether, Emmy, 131–2

Nørgaard, Per, 29

Norton, Simon, 2, 180–2, 184, 185, 189, 202–3, 227

nuclear forces, 76–7

number, irrational, 30 rational, 30

Oberwolfach, 107, 177

octahedron, 5–9, 103–5

odd order theorem 114, *see also* Feit-Thompson theorem

Ogg, Andrew, 193–4, 198–9

O'Nan, Michael, 180

Pacioli, Luca, 28

packing, 144–8

Parker, Richard, 182

periodic table, 66

permutation, 32 even, 36–40 odd, 36–8

Plato, 5

Platonic solids, 5–6

puzzle, 36–8

Pythagoras, 6

Pythagoreans, 6, 30, 224

quantum theory, 74, 76–7, 220, 222–4

quasi-thin, 212–15

Raspail, François-Vincent, 24

Ree, Rimhak, 95

regular polyhedron 103, *see also* Platonic solid

relativity theory, 72, 220–3

relativity, general, 73, 222

relativity, special, 72, 222

Renaissance, 17, 28

replication formulas, 203, 227

Revision (of the Classification), 211–13

Rubik cube, 49

Rudvalis, Arunas, 179

Ruffini, Paolo, 19–20

Schur, Issai, 116

Seitz, Gary, 167

Serre, Jean-Pierre, 194, 226

Shannon, Claude, 142

Shult, Ernest, 166
Sims, Charles, 139, 188
Smith, Stephen, 204, 214
Solomon, Ronald, 173, 211, 212
space-time, 73, 221–2, 225
Steinberg, Robert, 95, 168–9
string theory, 222–3, 225
Suzuki, Michio, 95, 122, 155
Swinnerton-Dyer, Sir Peter, 191
symmetry, 10
symmetry atom, *see* atom of
    symmetry

Tartaglia, 17–19
tesseract, 105–6
tetrahedron, 5–6, 103–4
Theaetetus, 5–6
Thompson, John, 1–2, 114, 121–7,
    135, 149–55, 184, 191–3,
    201–3
Thorne, Michael, 186–8

time-distance, 221, 223–4
Timmesfeld, Franz, 168
Tits, Jacques, 95, 98–9, 106–8,
    138, 209
topology, 198
torus, 198
torus, double, 198
transitivity, 128
    multiple, 129–31
transposition, 37, 160–1
transposition, Fischer, 161–70

Wales, David, 179
Walter, John, 173
Weil, André, 92–93
Weyl, Hermann, 117
Wilson, Robert, 182–3
Witt design, 133, 148, 240–1
Witt, Ernst, 131–3

Yeats, William Butler, 51